U.S. Department of Transportation
National Highway Traffic Safety Administration

DOT HS 811 304 April 2010

Quieter Cars and the Safety Of Blind Pedestrians: Phase I

Notice

This document is disseminated under the sponsorship of the Department of Transportation in the interest of information exchange. The United States Government assumes no liability for its contents or use thereof.

Notice

The United States Government does not endorse products or manufacturers. Trade or manufacturers' names appear herein solely because they are considered essential to the objective of this report.

REPORT DOCUMENTATION PAGE

Form Approved
OMB No. 0704-0188

Public reporting burden for this collection of information is estimated to average 1 hour per response, including the time for reviewing instructions, searching existing data sources, gathering and maintaining the data needed, and completing and reviewing the collection of information. Send comments regarding this burden estimate or any other aspect of this collection of information, including suggestions for reducing this burden, to Washington Headquarters Services, Directorate for Information Operations and Reports, 1215 Jefferson Davis Highway, Suite 1204, Arlington, VA 22202-4302, and to the Office of Management and Budget, Paperwork Reduction Project (0704-0188), Washington, DC 20503.

1. AGENCY USE ONLY (Leave blank) DOT HS 811 304	2. REPORT DATE April 2010	3. REPORT TYPE AND DATES COVERED Draft Report Final Report: April 2009 to January 2010
4. TITLE AND SUBTITLE Quieter Cars and the Safety of Blind Pedestrians: Phase I		5. FUNDING NUMBERS HS59A/HS76
6. AUTHOR(S) Lisandra Garay-Vega, Aaron Hastings, John K. Pollard, Michael Zuschlag, and Mary D. Stearns		
7. PERFORMING ORGANIZATION NAME(S) AND ADDRESS(ES) U.S. Department of Transportation Research and Special Programs Administration John A. Volpe National Transportation Systems Center Cambridge, MA. 02142-1093		8. PERFORMING ORGANIZATION REPORT NUMBER DOT-VNTSC-NHTSA-11-03
9. SPONSORING/MONITORING AGENCY NAME(S) AND ADDRESS(ES) U.S. Department of Transportation National Highway Traffic Safety Administration Human Factors Engineering Integration Division Vehicle Safety Research Office, NVS-331 Washington, DC 20590		10. SPONSORING/MONITORING AGENCY REPORT NUMBER DOT HS 811 304

11. SUPPLEMENTARY NOTES
1 = Volpe National Transportation Systems Center, RITA, US DOT

12a. DISTRIBUTION/AVAILABILITY STATEMENT Document is available to the public from the National Technical Information Service www.ntis.gov	12b. DISTRIBUTION CODE

13. ABSTRACT (Maximum 200 words)
The National Highway Traffic Safety Administration recognizes that quieter cars such as hybrid-electric vehicles in low-speed operation using their electric motors, may introduce a safety issue for pedestrians who are blind. This study documents the overall sound levels and general spectral content for a selection of hybrid-electric and internal combustion vehicles in different operating conditions, evaluates vehicle detectability for two ambient sound levels, and considers countermeasure concepts that are categorized as vehicle-based, infrastructure-based, and systems requiring vehicle-pedestrian communications. Overall sound levels for the hybrid-electric vehicles tested are lower at low speeds than for the internal combustion engine vehicles tested. There were significant differences in human subjects' response time depending on whether electric or internal combustion propulsion was used at both the lower and higher levels of ambient sound. Candidate countermeasures are discussed in terms of types of information provided (direction, vehicle speed, and rate of speed change, etc); useful range of detection of vehicles by pedestrians, warning time, user acceptability, and barriers to implementation. This study provides baseline data on the acoustic characteristics and auditory detectability of vehicles; however, the results cannot be generalized to more complex environments, as for example when multiple target vehicles are present.

14. SUBJECT TERMS Hybrid-electric vehicle, blind pedestrians, acoustic measurement, detectability		15. NUMBER OF PAGES 151
		16. PRICE CODE
17. SECURITY CLASSIFICATION OF REPORT Unclassified	18. SECURITY CLASSIFICATION OF THIS PAGE Unclassified	19. SECURITY CLASSIFICATION OF ABSTRACT Unclassified	20. LMITATION OF ABSTRACT Unlimited

NSN 7540-01-280-5500

Standard Form 298 (Rev. 2-89)
Prescribed by ANSI Std. 239-18
298-102

ACKNOWLEDGMENTS

The authors of this study are *Lisandra Garay-Vega, Aaron Hastings, John K. Pollard, Michael Zuschlag,* and *Mary D. Stearns* of the John A. Volpe National Transportation Systems Center (Volpe Center), Research and Innovative Technology Administration, United States Department of Transportation.

The authors acknowledge the contribution of many people. Special appreciation is due to the National Highway Traffic Safety Administration, including *Tim Johnson*, director of the Office of Human-Vehicle Performance Research; *Michael Perel*, former chief of the Human Factors/Engineering Integration Division and *Stephen Beretzky*, project manager in the Human Factors/Engineering Integration Division; as well as *Refaat Hanna* of NHTSA's National Center for Statistical Analysis; *Gayle Dalrymple*, NHTSA's Rulemaking; and *Paul Grygier, Riley Garrott,* and *Frank Barickman* of NHTSA's Vehicle Research and Test Center.

Special thanks to *Steven Rothstein,* president of the Perkins School for the Blind, Watertown, Massachusetts, for his willingness to collaborate in this study. Rothstein and the orientation and mobility specialists and staff provided information essential to characterizing the safety threat from quieter cars to pedestrians who are blind. Special appreciation is due to *Kim Charlson,* director of the Perkins Braille and Talking Book Library. She provided feedback on the study protocol as well as how to best distribute the recruitment information to the community. We also appreciate the assistance of *Robert Pierson,* recording studio director, Perkins Braille and Talking Book Library, for accommodating the team during the data collection period at the recording studio. We express gratitude to *Cynthia Essex*, the secondary program education director; to *Jan Seymour-Ford*, research librarian, and to members of the Perkins School community who participated in the study.

We thank *Arthur O'Neill*, vice president of the Carroll Center for the Blind, Newton, Massachusetts, for his willingness to collaborate in this study. We would like also to thank *Rabih Dow*, director of the Carroll Center's Residential Rehabilitation Training Program, and the orientation and mobility specialists. Meetings with the orientation and mobility specialists provided a valuable opportunity to help us understand the training process as well as the safety threats from quieter cars to pedestrians who are blind. *Joe Kolb* and *Maria de Los Angeles Goldstein* discussed and demonstrated the information-gathering process and strategies typically used by pedestrians who are blind. Joe Kolb also provided valuable guidance and feedback on the informed consent form and study procedures. *Heather Platt* provided references to relevant study material and hosted our team when they conducted acoustic measurements on the Carroll Center campus. We also appreciate the assistance of *Padma Rajagopal* in recruiting and scheduling participants during the data collection period. We express gratitude to members of the Carroll Center community who participated in the study.

The authors acknowledge the feedback and information exchange with researchers and representatives of several organizations: *Debbie Stein,* chairperson of the National Federation of the Blind Committee on Automobiles and Pedestrians; *Eric Bridges,* director of Advocacy & Governmental Affairs, American Council of the Blind; *Meg Robertson,* director of orientation and mobility, Department of the Massachusetts Commission for the Blind; *Lawrence D. Rosenblum,* professor at the University of California-Riverside; *Robert Wall Emerson,* professor at Western Michigan University; the *Alliance of Automobile Manufacturers*; the *Society of Automotive Engineers Vehicle Sound for Pedestrian Committee*; and the *Association of International Automobile Manufacturers.*

The authors acknowledge the following Volpe Center staff who made valuable contributions to this study: *John D. Smith* for the information and feedback provided throughout the study; *Christopher Scarpone* and *Robert Samiljan* for their assistance during the acoustic recordings; *Matthew Isaacs* for consultation on experimental setup and data management; and *Catherine Guthy* for her assistance during the human-subject studies. We thank *Stephen Popkin,* director of the Human Factors Center of Innovation and *Gregg Fleming,* director of the Environmental and Energy Systems Center of Innovation for their support. We also appreciate the support of *Christopher Roof,* chief of the Environmental Measurement and Modeling Division.

Finally, the authors extend special appreciation to the volunteers who participated in the human subject study.

LIST OF ABBREVIATIONS AND ACRONYMS

AASHTO	American Association of State Highway and Transportation Officials
ACB	American Council of the Blind
AFB	American Foundation for the Blind
AIAM	Association of International Automobile Manufacturers
ANC	Active noise cancellation
ANOVA	Analysis of variance
Alliance	Alliance of Automobile Manufacturers
CANbus	Controller Area Network bus
dB	Decibel
dB(A)	A-weighted decibel level
DSP	Digital signal processing
EM	Electric mode
EVs	Electric vehicles
GPS	Global positioning system
HEVs	Hybrid-electric vehicles
ICE	Internal combustion engine
IRB	Institutional Review Board
JAMA	Japan Automobile Manufacturers Association
JASIC	Japanese Automobile Standards Internationalization Centre
$L_{A(eq)}$	Average A-weighted sound level
$L_{A(max)}$	Maximum A-weighted sound level
$L_{A(min)}$	Minimum A-weighted sound level
LD824	Larson Davis Model 824
mph	Miles per hour
NCSA	National Center for Statistics and Analysis
NHTSA	National Highway Traffic Safety Administration
NFB	National Federation of the Blind
OEM	Original equipment manufacturer
RITA	Research and Innovative Technology Administration
SAE	Society of Automotive Engineers

SDS	State data system
SLM	Sound level meter
SPL	Sound pressure level
TRC	Transportation Research Center
USDOT	United States Department of Transportation
UTC	Universal Time Code
VIN	Vehicle identification number
Volpe Center	Volpe National Transportation Systems Center
VRTC	Vehicle Research and Test Center
VSP	Vehicle Sound for Pedestrian Subcommittee

TABLE OF CONTENTS

EXECUTIVE SUMMARY 1
1. INTRODUCTION 6
    1.1 ISSUE 6
    1.2 PROJECT BACKGROUND 6
    1.3 GOALS AND OBJECTIVES 7
    1.4 SCOPE 7
    1.5 STUDY APPROACH 8
    1.6 REPORT ORGANIZATION 9
2. LITERATURE REVIEW 10
    2.1 CRASH INCIDENCE: HYBRID VERSUS NON-HYBRID VEHICLES 10
        2.1.1 Vehicle Maneuver Prior to Crash 11
        2.1.2 Speed Limit 11
        2.1.3 Lighting and Weather Conditions 11
    2.2 ANECDOTAL REPORTS 11
    2.3 BLIND PEDESTRIAN MOBILITY NEEDS AND ACOUSTIC CUES FOR TRAVEL 12
        2.3.1 Information and Strategies Used by Pedestrians Who Are Blind 12
        2.3.2 Conflicts at Sites Without Traffic Control 13
        2.3.3 Conflicts at Stop-Controlled Intersections 14
        2.3.4 Conflicts at Signalized Intersections 14
        2.3.5 Complex Non-Controlled Intersections 14
        2.3.6 Parking Lots and Driveways 15
    2.4 VEHICLE OPERATION, TRAFFIC SOUNDS, AND VEHICLE DETECTABILITY 15
        2.4.1 Vehicle Approaching at Low Speed 17
        2.4.2 Vehicle Stationary 19
        2.4.3 Vehicles Accelerating From a Stop 19
    2.5 DRIVER REACTION TIME, BRAKING DISTANCE, AND PEDESTRIAN SAFETY 20
3. CRITICAL SAFETY SCENARIOS FOR PEDESTRIANS WHO ARE BLIND 21
    3.1 VEHICLE BACKING OUT 21
    3.2 VEHICLE TRAVELING IN PARALLEL AND SLOWING 21
    3.3 VEHICLE APPROACHING AT LOW SPEED 22
    3.4 VEHICLE ACCELERATING FROM A STOP 22
    3.5 VEHICLE STATIONARY 22
    3.6 AMBIENT SOUND LEVELS 23
4. ACOUSTIC MEASUREMENT OF VEHICLES AND AMBIENT SOUNDS 24
    4.1 PROCEDURE TO MEASURE VEHICLES AND AMBIENT SOUNDS 24
        4.1.1 Objectives 24
        4.1.2 Test Vehicles 24
        4.1.3 Vehicle Measurement Site 25
        4.1.4 Instrumentation and Equipment Layout 28

        4.1.5   Data Collection Protocol 29

        4.1.6   Data Collection Procedure for Each Vehicle Operation 31

   4.2   RESULTS OF ACOUSTIC MEASUREMENT OF VEHICLE AND AMBIENT SOUNDS .. 32

        4.2.1   Vehicle Acoustic Measurement Results 32

        4.2.2   Comparison of SAE 6.56-ft With the 12-ft Microphone Test 41

   4.3   AMBIENT MEASUREMENT FOR CRITICAL SAFETY SCENARIOS 42

   4.4   DISCUSSION 43

5. AUDITORY DETECTABILITY OF VEHICLES IN CRITICAL SAFETY SCENARIOS 45

   5.1   PROCEDURE TO MEASURE THE AUDITORY DETECTABILITY OF VEHICLES 45

        5.1.1   Objectives 45

        5.1.2   Subjects 45

        5.1.3   Apparatus 47

        5.1.4   Study Design and Methods 47

        5.1.5   Data Reduction 48

   5.2   RESULTS OF HUMAN SUBJECT STUDIES 49

        5.2.1   Vehicle Backing Out 50

        5.2.2   Vehicle Approaching in Parallel and Slowing 53

        5.2.3   Vehicle Approaching at a Constant Low Speed 57

   5.3   DISCUSSION 61

6. EXAMINATION OF POTENTIAL COUNTERMEASURES 63

   6.1   COUNTERMEASURE ALTERNATIVES 63

        6.1.1   Vehicle-Based Audible Alert Signals 66

        6.1.2   Systems Requiring Vehicle-Pedestrian Communications 67

   6.2   ADVANTAGES AND DISADVANTAGES OF POTENTIAL COUNTERMEASURES 68

   6.3   DISCUSSION 68

7. SUMMARY OF FINDINGS AND CONSIDERATIONS FOR FURTHER RESEARCH 71

   7.1   FINDINGS 71

   7.2   CONSIDERATIONS FOR FURTHER RESEARCH 73

8. BIBLIOGRAPHY 75

APPENDIX A.   ACOUSTIC DATA FOR VEHICLES A-1

LIST OF TABLES

Table 1. Vehicles Included in the NCSA Analysis 10
Table 2. Sound Intensity by Vehicle Operating Condition (Wiener & Lawson, 1997) 16
Table 3. Sound Intensity of Vehicles Approaching at 30 mph (Wiener, Naghshineh,
 Salisbury, & Rozema (2006) 17
Table 4. Test Conditions in Wiener, Naghshineh, Salisbury, and Rozema (2006) 19
Table 5. Test Vehicles, Make, Model Year and Engine Type 24
Table 6. Passby Levels for Reverse Constant Speed Passby Events at the
 12-ft Microphone Location 35
Table 7. Passby Levels for Deceleration Passby Events at the 12-ft Microphone Location 35
Table 8. Passby Levels for 6 mph Constant Speed Passby Events at the
 12-ft Microphone Location 36
Table 9. Overall Levels by Vehicle for Idle at the 12-ft Microphone Location 36
Table 10. Subject Age Distribution 46
Table 11. Subject Vision Loss Category 46
Table 12. Subject Mobility Aid Usage 46
Table 13. Test Conditions for Human Subject Studies 47
Table 14. Mean Response Time and Time-to-Vehicle-Arrival: Vehicles Backing Out 52
Table 15. Mean Response Time and Time-to-Vehicle-Arrival: Vehicle Slowing 56
Table 16. Mean Response Time and Time-to-Vehicle-Arrival: Vehicle Approaching at Low
 Speed 58
Table 17. Mean Detection Distance for Vehicle Approaching at a Constant Low Speed
 (Low Ambient Sound) 60
Table 18. Mean Detection Distance for Vehicle Approaching at a Constant Low Speed
 (High Ambient Sound) 61
Table 19 Average Times to Vehicle Arrivals (seconds) 61
Table 20. Pedestrian Safety Countermeasures 64

LIST OF FIGURES

Figure 1. Tire Tread Showing Debris Removed Prior to Testing 25
Figure 2. Aerial View of Vehicle Measurement Site 26
Figure 3. Vehicle Measurement Sub-Site 1 Facing Southeast 27
Figure 4. Vehicle Measurement Sub-Site 1 Facing Northeast 27
Figure 5. Vehicle Measurement Equipment Layout, Sub-Site 1 28
Figure 6. Vehicle Measurement Equipment Layout, Sub-Site 2 29
Figure 7. Sound Pressure Level in dB(A) for Prius 20 mph Passby (No Correction Needed) 33
Figure 8. Sound Pressure Level in dB(A) for Prius 10 mph Passby (Suitable for Correction) 34
Figure 9. Sound Pressure Level in dB(A) for Prius Idle
 (Too Close to Background for Correction) 34

Figure 10. Sample One-Third Octave Band Spectra for Reverse 5 mph Constant Speed Passby at 12 ft Microphone Location 37

Figure 11. Sample One-Third Octave Band Spectra for Decelerating Passby at 12-ft Microphone Location 38

Figure 12. Sample One-Third Octave Band Spectra for 6 mph Constant Speed Passby at 12-ft Microphone Location 39

Figure 13. Maximum Levels in dB(A) for the Prius (O) and Matrix (X)..................................40

Figure 14. Maximum Levels in dB(A) for the Civic Hybrid (O) and ICE (X)...........................40

Figure 15. Maximum Levels in dB(A) for the Highlander Hybrid (O) and ICE (X) 41

Figure 16. Comparison of Idle Measurements at 6.6 ft (according to SAE J2889-1) and 12 ft (Volpe)...42

Figure 17. Average A-Weighted One-Third Octave Band Levels for Ambient Measurements: All Sites 43

Figure 18. Missed Detection Rates for the "Vehicle Backing Out" Scenario 50

Figure 19. Distribution of Subjects by the Number of Missed Detection: Vehicles Backing Out 51

Figure 20. Mean Time-to-Vehicle-Arrival by Vehicle and Ambient Condition: Vehicles Backing Out 52

Figure 21. Narrow-band Spectral Analysis of Prius Braking Sound 53

Figure 22. Missed Detection Rates for the "Vehicle Approaching in Parallel and Slowing" Scenario 54

Figure 23. Distribution of Subjects by the Number of Missed Detection: Vehicle Slowing..........55

Figure 24. Mean Time-to-Vehicle-Arrival: Vehicle Slowing 56

Figure 25. Missed Detection Rates for the "Vehicle Approaching at a Constant Speed" Scenario 57

Figure 26. Distribution of Subjects by the Number of Missed Detection: Low Speed 58

Executive Summary

PURPOSE

The National Highway Traffic Safety Administration's Office of Vehicle Safety Research tasked the Volpe National Transportation Systems Center of the U.S. Department of Transportation's Research and Innovative Technology Administration to examine the issue of quieter cars and the safety of pedestrians who are blind. Quieter cars such as electric vehicles (EVs) and hybrid-electric vehicles (HEVs) can reduce pedestrians' ability to assess the state of nearby traffic and, as a result, may have an adverse impact on pedestrian safety. The primary concern is when HEVs operate using their electric motor systems at slow speeds when other auditory cues from tires and wind noise are less dominant. A perceived reduction in the sound emitted by these vehicles creates a safety concern because these sounds are often the best or only source of information that pedestrians who are blind use to avoid conflicts. A significant reduction in auditory cues from vehicles may impact the ability of pedestrians who are blind to travel safely. This study examines the acoustic characteristics of a selection of HEVs and internal combustion engine (ICE) vehicles, assesses their auditory detectability in various operations and ambient sound conditions, and discusses potential countermeasures.

METHODS

This study describes critical safety scenarios considered in the evaluation and comparison of the acoustic characteristics of a selection of vehicles and the auditory detectability of these vehicles by pedestrians who are blind. These scenarios consider the safety risks, information needs, and strategies used by pedestrians who are blind. The scenarios identified include vehicles backing out, vehicles turning right into the pedestrian's path, vehicles approaching at a constant speed, vehicles accelerating from a stop, and stationary vehicles that could suddenly move. They were defined by combining pedestrian-vehicle environments, vehicle maneuvers/speeds/operations, and ambient sound levels. The scenario criteria were compiled from the following activities: review of crash data analyses and anecdotal reports; literature reviews; observations of how blind pedestrians are trained to navigate in various pedestrian-vehicle environments; and cognitive walkthroughs where pedestrians who are blind follow cues and strategies from orientation and mobility specialists at various pedestrian-vehicle environments. Pedestrians who are blind use acoustic cues from vehicles to get information about vehicle presence, vehicle position relative to the pedestrian, vehicle direction of travel, and vehicle rate of acceleration or speed. Traffic sounds also provide cues that help them to orient towards the crosswalks, identify a time to cross, and maintain alignment while crossing. This study identifies the following operating conditions that mimic vehicles operating in a critical safety scenario: vehicle backing out at 5 mph (mimicking a vehicle backing out of a driveway); vehicle slowing from 20 to 10 mph (mimicking a vehicle preparing to turn right from the parallel street); vehicle approaching at a constant low speed; vehicle accelerating from a stop; and vehicle stationary (such as at a stop light).

This study also describes the digital recordings and acoustic measurements collected to document the sound emitted by HEVs and ICE vehicles in the operating conditions described above. Acoustic data for vehicles approaching at low speeds (6 mph and 10 mph) and moderate speeds (20 mph, 30 mph, 40 mph) were recorded to document how the overall sound level for ICE vehicles and their HE twins differ as a function of vehicle speed. The vehicles used in the study were: the Toyota Prius and Toyota Matrix (a proxy for an ICE twin due to similar size and weight); the Honda Civic Hybrid and Honda Civic ICE; and the Toyota Highlander Hybrid and Toyota Highlander ICE. The vehicle measurements were conducted at NHTSA's Vehicle Research and Test Center located at the Transportation Research Center in East Liberty, Ohio. Most of the data was recorded during the night (with no other vehicle operating at the test facility) to guarantee the quietest possible background levels at the site and obtain the highest quality recordings possible. The average A-weighted sound pressure level at the test site during the recordings was 31.2 dB(A). The measurement procedure used follows recommendations of the Society of Automotive Engineers draft test procedure for "Measurement of Minimum Noise Emitted by Road Vehicles" but deviates somewhat because the goal of this study is to document vehicle acoustics under "critical safety scenarios" rather than just the "minimum noise emitted," as is the case with the SAE document. For example, the study includes more vehicle operating conditions than specified in the SAE draft test procedure and uses additional microphones placed in positions corresponding to the anatomical location of human ears. Acoustic measurements were made for a selection of stationary vehicles using the SAE method to document the differences between the two approaches. Ambient sound levels for representative geographic locations where pedestrians could expect to hear a nearby vehicle were recorded and used to examine vehicle detectability at two ambient sound levels.

Human subject studies were carried out to examine the auditory detectability of four vehicles (two ICE vehicles and two HEVs operated in electric mode) in three vehicle operating conditions for two ambient sound levels. The three operating conditions used for the human-subject studies were: vehicle backing out at 5 mph (mimicking a vehicle backing out of a driveway); vehicle slowing from 20 to 10 mph (mimicking a vehicle preparing to turn right from the parallel street); and vehicle approaching at a constant speed (6 mph). The two ambient sound levels considered in the human subject studies were a relatively quiet rural ambient with overall sound level of 31.2 dB(A) and moderately noisy suburban ambient sound with overall sound level of 49.8 dB(A). The study excluded simulations of noisier environments because preliminary data and previous studies suggested that the differences in detectability between the two types of vehicles are small when ambient sound levels are high — both types of vehicles become difficult to detect. The human subject studies were completed in a laboratory setting and subjects were legally blind, self-reported to have normal hearing, independent travelers, and 18 or older.

Last, this study identified countermeasure concepts from literature reviews. These concepts are categorized as vehicle-based, infrastructure-based, and systems requiring vehicle-pedestrian communications. These concepts were reviewed and compared considering the information provided, potential for user acceptance, and implementation.

RESULTS

This study examined how the overall sound levels for ICE vehicles and HEVs differ as a function of vehicle speed and maneuver. Vehicles sound levels were measured when stationary, backing out, moving at a constant speed and slowing, accelerating from a stop, and approaching at low and moderate speeds. The overall sound levels for the HEVs tested are lower than for the ICE vehicles tested. The overall sound levels for the Toyota hybrids when stationary were too low to be recorded under the ambient conditions present. The sound levels for the two types of vehicles converge at higher speeds; maximum differences were noted at 5 or 6 mph, with much smaller differences at 10 mph and no significant difference after 20 mph. The overall sound levels when backing out are clearly lower for the HEVs tested than for the ICE vehicles tested (7 to 10 dB(A) difference). On the other hand, for vehicles slowing, the difference between HEVs and ICE vehicles tested is small (2 dB(A)). It is important to mention that the Toyota HEVs tested emitted a faint 5-kHz tone with a 10-kHz harmonic when they were slowing or braking. This tone is associated with the electronic components of the vehicle when braking (e.g., regenerative braking); it is not intended to warn pedestrians. The difference in sound levels between the HEVs and ICE vehicles tested is also small when vehicles accelerate from a stop. The overall sound levels for a vehicle approaching at a low speed (6 mph) are clearly lower for the HEVs tested than for the ICE vehicles tested (2 to 8 dB(A)). Sounds at the same level appear to be more or less detectable depending on their spectral shape. Considering the one-third octave band spectrum, there is a tendency for the HEVs tested to have less high-frequency content relative to the overall sound level—an exception to this is the notable 5-kHz peak in the Toyota HEVs when slowing or braking.

The human-subject studies collected data on whether subjects can detect a target vehicle and how soon they can detect it before the vehicle reaches their position (time-to-vehicle-arrival is the metric for pedestrian performance) for two ICE vehicles (Toyota Matrix and Highlander) and two HEVs (Toyota Prius and Highlander) in electric-vehicle mode. These studies also examined the effect of ambient sound levels on the ability of blind subjects to detect vehicles. Subjects listened to audio recordings of three vehicle maneuvers: vehicle backing out on the left side as if from a driveway; vehicle approaching at a constant low speed of 6 mph from the left; and vehicle moving parallel to the pedestrian and slowing from 20 mph to 10 mph as if to turn right. Overall, subjects were generally able to detect a target vehicle present in a given scenario: 95.8 percent of the subjects detected a vehicle approaching at 6 mph, 89.6 percent detected a vehicle backing out at 5 mph, and 83.3 percent detected a vehicle slowing from 20 to 10 mph. Two subjects (4.2%) never detected a Toyota Prius approaching at 6 mph. Five subjects (10.4%) never detected one or more HEVs backing out. Eight subjects (16.7%) never detected slowing vehicles (most frequently the Toyota Matrix). Average times-to-vehicle-arrival across all subjects are summarized in Table ES-1 for the vehicles and ambient sound level conditions tested. Time-to-vehicle-arrival is the time from first detection of a target vehicle to the instant the vehicle passes the microphone line/pedestrian location.

Table ES-1. Average Times-to-Vehicle-Arrivals (seconds)

Vehicle Maneuver	Ambient Sound Level			
	Low		High	
	HEVs	ICE Vehicles	HEVs	ICE Vehicles
Backing out (5 mph)	3.7	5.2	2.0	3.5
Slowing from 20 to 10 mph	2.5	1.3	2.3	1.1
Approaching at 6 mph	4.8	6.2	3.3	5.5

In nearly all cases, subjects detect vehicles later (i.e., closer to vehicle arrival at the pedestrian position) in the high ambient sound condition than in the low ambient sound condition. Time-to-vehicle-arrival is shorter for the HEVs tested than for the ICE vehicles tested, except for the slowing scenario. The differences in time-to-vehicle-arrival between HEVs tested and their ICE twins are statistically significant within a vehicle maneuver and ambient sound level condition. Difference for HEVs backing out versus HEVs slowing in the high ambient condition are not significant. The anomalous result in the case of slowing vehicles (i.e., HEVs are detected sooner) is attributed to the 5-kHz tone emitted by the Toyota HEVs, which is loudest when they are in regenerative braking mode.

DISCUSSION

The results of this study show that HEV models tested differ from ICE vehicles in operation, sound levels emitted, and spectral content. Although the HEVs tested were detected later than the ICE vehicles (except for slowing vehicles), the times at which the subjects detected the vehicle would usually be sufficient for the pedestrian or the driver to take evasive action. The study mimics the situation in which a blind pedestrian knows there is a high probability of hearing a vehicle within a few seconds and can devote full attention to listening for it. It is reasonable to expect that the response times observed will be longer in normal walking situations when the demand for pedestrians' attention is higher. This study examined situations when one target vehicle was present and there were either a few vehicles in the background or they were several blocks away from the pedestrian location. The results provide baseline data on the acoustic characteristic and auditory detectability of vehicles; however, the results cannot be generalized to more complex environments; for example, when multiple target vehicles are present.

Preliminary evaluation criteria to compare countermeasures include: types of information provided (direction, vehicle speed, and rate of speed change, etc.); pedestrian detection range, warning time, user acceptability, and barriers to implementation. Infrastructure options and pedestrian training can improve safety; however, they cannot directly address the issue that HEVs operated in electric mode are not detectable in many situations in which conventional vehicles are clearly audible. At the present time, only countermeasures that generate alert sounds emulating those of approaching ICE vehicles come close to meeting the requirements of pedestrians who are blind. Other countermeasures fail to provide sufficient cues about vehicle position, speed, and rate of change in speed. Sound content, such as the relative proportions of high and low frequencies, can be manipulated to improve the effectiveness of such alert sounds while reducing the overall community noise impact. Considering the results presented in this study, such sounds are only needed when vehicles are operated at very low speeds (less than 20 mph). Groups representing people who are blind have expressed a preference for a sound that mimics the sound characteristics of an ICE vehicle in different operating conditions (e.g., low speed, acceleration/deceleration, stationary). This would make recognition of the alert sound intuitive to all pedestrians. The characteristic sound of an ICE being started is often the first cue of the presence of a potential threat in a parking lot; it is desirable that this vehicle state is also represented on HEVs or EVs.

This study documents the overall sound levels and general spectral content for six vehicles in different operating conditions. Vehicle detectability was evaluated in two ambient sound levels. Follow-on research needs to consider evaluating the effectiveness and user acceptance of vehicle-emitted sound to alert blind pedestrians to vehicle presence, direction, location, and operation. For example, it is necessary to evaluate the recognition, response time, and accuracy in a complex soundscape of sources that emulate slow-moving vehicles all around the subject. Further evaluation is needed to define what kinds of synthetic sound (e.g., broadband, tonal, modulation, or a combination) provide the most useful alert cues for pedestrians.

1. INTRODUCTION

1.1 Issue

Quieter cars such as electric vehicles and hybrid-electric vehicles can reduce pedestrians' ability to assess the state of nearby traffic and, as a result, can have an impact on pedestrian safety. The primary concern is when HEVs operate using their electric motor system at slow speeds when other auditory cues from tires and wind noise are less dominant. EVs and HEVs can have a beneficial outcome for environmental and noise reduction but they may pose a safety problem for pedestrians, in particular blind pedestrians, who rely on auditory cues from vehicles to navigate. Vehicle type, vehicle speed, and ambient sound levels are some of the factors that could influence the auditory detectability of vehicles. This report describes the first phase of the National Highway Traffic Safety Administration's study to examine the safety risk associated with quieter cars for pedestrians who are blind and to investigate appropriate countermeasures.

A significant reduction in auditory cues from vehicles may reduce the available information needed by blind and other vision-impaired pedestrians to navigate, and thus impact their ability to travel safely. Blind pedestrians use auditory information to determine the position of the vehicle relative to themselves, the direction of travel, and the rate of acceleration or vehicle speed. Auditory cues from vehicles also facilitate pedestrian orientation tasks such as establishing alignment before crossing and while in the crosswalk.

About 3.3 million Americans 40 and older are blind or have low vision; this number is estimated to increase 70 percent, reaching 5.5 million in the year 2020 (Eye Diseases Prevalence Research Group, 2004). The magnitude and details of the impact of quieter cars on the safety of pedestrians who are blind is not well known. This study describes studies completed by the John A. Volpe National Transportation Systems Center of the Research and Innovative Technology Administration, United States Department of Transportation, that examine the issue of quieter cars on the basis of vehicle type, vehicle operation, and ambient sound levels. The study also presents a review of potential countermeasures to mitigate the safety risk and identifies considerations for future evaluations.

1.2 Project Background

In December 2007, NHTSA met with representatives of the blind community to discuss the issue of quieter cars and the safety of pedestrians who are blind. Since 2007 NHTSA has been monitoring the work of the Society of Automotive Engineers, Vehicle Sound for Pedestrian Subcommittee to the SAE Safety and Human Factors Committee to stay abreast of developments concerning the issue of quieter cars. The VSP subcommittee includes members of the Alliance of Automobile Manufacturers (the Alliance), the Association of International Automobile Manufacturers, and the SAE Safety and Human Factors Committee, among others. In June 2008, NHTSA held a public meeting to provide a forum for interested parties to discuss the issue of quieter cars. NHTSA established a docket to collect information on the issue and developed a research plan ("Quieter Cars and the Safety of Blind Pedestrians: A Research Plan," April 2009) to investigate the safety risk to blind pedestrians posed by quieter cars. The final research plan was posted in the docket NHTSA-2008-0108 on May 6, 2009. NHTSA selected the Volpe Center in Cambridge, Massachusetts, to conduct this research due to their expertise in acoustics measurements and human factors research.

1.3 Goals and Objectives

The goals of this study are to record and compare the acoustic parameters of a selection of HEVs and ICE vehicles in various operating conditions, and to examine the auditory detectability of these vehicles in various ambient sound levels. The study also reviews possible countermeasures to address the quieter cars issue, and discusses considerations for future evaluations of countermeasures.

The following objectives were established to meet the goals of the study:

1. Characterize the safety problem;
2. Identify requirements for blind pedestrians' safe mobility (emphasizing acoustic cues from vehicles and ambient sound); and
3. Identify potential countermeasures and describe their advantages and disadvantages.

The following key tasks were completed to address the objectives:

1. Identify/define critical safety scenarios where pedestrian-vehicle conflicts are likely to occur and describe potential contributing factors.
2. Describe the information used by blind pedestrians, how the information is perceived, and how a reduction of auditory cues from vehicles may impact pedestrians' safety.
3. Review the SAE J2889-1 draft test procedure for acoustic measurements of vehicles. Develop a procedure incorporating usable aspects of the SAE procedure to measure the acoustic characteristics of HEVs and ICE vehicles in various operating conditions (corresponding to critical safety scenarios).
4. Measure and record the acoustic characteristics of HEVs and ICE vehicles in various operating conditions; record ambient sound levels; combine and reproduce the sounds of vehicles and ambient sounds for use in human-subject studies.
5. Conduct human subject studies to examine the auditory detectability of HEVs and ICE vehicles in critical safety scenarios.
6. Identify potential countermeasures and criteria to evaluate them.
7. Review the strengths and limitations of potential countermeasures.

1.4 Scope

This final report describes the results of the 10-month study to examine the acoustic characteristics of vehicles and the risk to blind pedestrians from quieter cars. It addresses four topic areas: (1) definition of critical safety scenarios based on contributing factors and auditory information used by blind pedestrians to travel; (2) documentation of the acoustic characteristics for vehicles and ambient sounds; (3) examination of the auditory detectability of vehicles in critical safety scenarios; and (4) identification of criteria to evaluate potential countermeasures, and a review of these countermeasures.

Acoustic measurements include data for three HEVs and three ICE vehicles in six operating conditions. These were conducted at a closed-course facility with all sources of ambient noise minimized. The auditory detectability of two HEVs and two ICE vehicles was evaluated in three operating conditions (corresponding to critical safety scenarios) selected from the six operating conditions used for the acoustic recordings. Human subject studies were completed in a laboratory setting and included data for 48 subjects. Subjects were legally blind, self-reported to have normal hearing in both ears (without hearing aids), traveled independently on a regular basis, and were at least 18 years old. The discussion on potential countermeasures is based on literature reviews, and considers information gathered from acoustic data and human subject studies. Testing of potential countermeasures is not within the scope of this study.

1.5 Study Approach

Volpe Center activities are grouped in four areas corresponding to the topic areas described above.

The first area, *definition of critical safety scenarios*, includes a review of crash data involving pedestrians and HEVs as reported by NHTSA's National Center for Statistics and Analysis and anecdotal events gathered by NFB involving blind pedestrians. In addition, the Volpe Center completed cognitive walkthroughs with pedestrians who are blind and orientation and mobility specialists, and participated in meetings with interested parties to discuss the issue. The Volpe Center also identified, obtained, and reviewed research related to orientation and mobility for blind pedestrians, vehicle sounds, and the auditory detectability of vehicles. Information from these activities was used to define the critical safety scenarios for evaluation. The term *critical safety scenario* describes the pedestrian-vehicle environments, vehicle operations, and ambient sound levels considered in the evaluation and comparison of the acoustic characteristics of a selection of vehicles and the auditory detectability of these vehicles by pedestrians who are blind.

The second area includes *acoustic measurements of HEVs and ICE vehicles and ambient sounds*, which include overall sound levels and frequency. The third area consists of *human subject studies conducted in a laboratory setting to examine response time and detection accuracy*. The information from these tasks was used to document the acoustic characteristics of a selection of vehicles and to examine the auditory detectability of these vehicles. The fourth area includes a review of *potential countermeasures (available or proposed) to reduce the risk from quieter cars*.

The Volpe Center reviewed the strengths and limitations of potential infrastructure-based, pedestrian-based, and vehicle-based countermeasures. In addition, the information obtained as a result of these activities was used to identify potential countermeasures that may merit future evaluation.

1.6 Report Organization

- Chapter 2 discusses the literature review topics: incidence rate of pedestrian crashes by HEVs and ICE vehicles; blind pedestrian mobility needs and acoustics cues used for travel; traffic sounds, and vehicle detectability. This information provides the foundation to define the critical safety scenarios.
- Chapter 3 discusses the critical safety scenarios and the mobility requirements and tasks of pedestrians who are blind.
- Chapter 4 discusses the measurements of the acoustic characteristics of a selection of vehicles in operating conditions corresponding to the critical scenarios and the measurements of ambient sounds.
- Chapter 5 describes the human subject testing, including pedestrians' response time, and accuracy in detecting vehicles in specific maneuvers and ambient sound levels.
- Chapter 6 reviews potential countermeasures, including their advantages and disadvantages.
- Chapter 7 discusses the summary of findings, considerations, and future research.

2. LITERATURE REVIEW

2.1 Crash Incidence: Hybrid Versus Non-Hybrid Vehicles

Crash data is one of the resources used in this study to identify the pedestrian-vehicle environments of interest (e.g., driveways, intersections) and vehicle maneuvers (e.g., turning, backing out into pedestrian path) to be considered in the evaluation of vehicle detectability. NHTSA's NCSA documented the incidence rate of pedestrian and bicyclist crashes with HEVs and compared the results to ICE vehicles. Incident rates are calculated as the number of vehicles (HEV or ICE) involved in crashes with pedestrians in a particular situation, divided by the total number of vehicles (HEV or ICE) that were in any crash under that same situation. The incidence rate of pedestrian crashes was found to be higher for HEVs than the ICE vehicles in the study, and the difference is statistically significant. A total of 8,387 HEVs and 559,703 ICE vehicles were included in the analysis. Seventy-seven HEVs and 3,578 ICE vehicles were involved in crashes with pedestrians. This figure represents 0.9 percent of all HEVs and 0.6 percent of all ICE vehicles in the analysis (Hanna, 2009).

The analysis by NCSA compares the crash incidence rate for the two types of vehicles when a pedestrian was the first event that caused an injury or damage in a crash. The analysis uses a small sample (8,387 HEVs and 559,703 ICE vehicles) and does not intend to provide national estimates. Files from the State Data System were used to compare the crash incidence by HEVs and ICE vehicles for model years 2000 and later. The SDS is maintained by NCSA and includes records for all crashes reported to the police regardless of the crash or injury outcome. The analysis was limited to data from the 12 States that include the vehicle identification number in the crash records. The VIN was used to identify vehicle make, model, and type (HEV or ICE). Table 1 shows the vehicles included in each of the two categories. Data availability years vary across the 12 States. The analysis includes an average of six years of crash data per State; data availability years range from 2000 to 2007. The SDS does not include information on pedestrian vision status; this study shows data for all pedestrians. Some of the situations included in the analysis are: (1) vehicle maneuver prior to the crash; (2) speed limit as a proxy for vehicle travel speed; and (3) weather and lighting condition at the time of the crash. Pedestrian crashes are described below for the two types of vehicles in each of the three situations.

Table 1. Vehicles Included in the NCSA Analysis

Hybrid Electric Vehicles (Total = 8,387)	Internal Combustion Engine Vehicles (Total = 559,703)
Toyota Corolla	Toyota Corolla
Toyota Camry	Toyota Camry
Honda Civic	Honda Civic
Honda Accord	Honda Accord
Toyota Prius	

2.1.1 Vehicle Maneuver Prior to Crash

NCSA examined pedestrian crashes for the two types of vehicles by vehicle maneuver prior to the crash. There is a significant difference in the crash incidence rate between HEVs (1.8%) and ICE vehicles (1.0%) for turning maneuvers. Of the 1,061 HEVs making a turn prior to a crash, 19 involved pedestrians as the first harmful event in the crash. Of the 70,245 ICE vehicles making a turn prior to a crash, 698 involved pedestrians as the first harmful event in the crash.

Pedestrian crashes where the vehicle is backing out, slowing/stopping, starting in traffic, and, entering or leaving a parking space/driveway, were combined and compared for the two types of vehicles. The crash incidence rate for the combined set of maneuvers is 1.2 percent and 0.6 percent for HEVs and ICE vehicles respectively; the difference (0.6%) is statistically significant. Of the 1,454 HEVs making these maneuvers prior to a crash, 17 involved pedestrians as the first harmful event in the crash. Of the 90,003 ICE making these maneuvers, 514 involved pedestrians as the first harmful event in the crash.

2.1.2 Speed Limit

The incidence of pedestrian crashes in zones with speed limits of 35 miles per hour (mph) or less is 0.5 percent higher for HEVs. In general, most pedestrian crashes occurred in zones with speed limit less than 35 mph. NCSA examined pedestrian vehicle crashes involving HEVs and ICE vehicles using speed limit as a proxy for vehicle speed. A total of 2,609 HEVs were involved in crashes in zones with speed limit less than 35 mph, and 48 of these involved pedestrians. A total of 152,833 ICE were involved in crashes in zones with speed limits less than 35 mph, and 1,836 of these involved pedestrians. The difference (0.6%) in incidence of pedestrian crashes in zones with speed limits of 35 mph or less between HEVs (1.8%) and ICE (1.2%) vehicles is statistically significant.

2.1.3 Lighting and Weather Conditions

In general, most pedestrian crashes occurred in daylight and during clear weather. The incidence of pedestrian crashes by HEVs is slightly higher than for ICE vehicles during dawn/dusk and when streets are dark (street lights off). Similarly, the crash incidence when raining, snowing, and cloudy/foggy is somewhat higher for HEVs. Sample size used in the NCSA analysis did not permit significance testing.

2.2 Anecdotal Reports

The NFB gathered subjective information from people involved in crashes or conflicts with vehicles. The information shows the vehicle maneuvers and pedestrian-vehicle environment of concern, which are consistent with the results of the NCSA crash data analysis. These concerns include vehicles turning, backing out, or moving into/from a driveway, alley, or parking lot.

The survey used by the NFB ("Survey on Quiet Cars Incidents") includes 46 questions. Twelve questions collect personal information; 9 questions gather background information on the incident; 12 questions are about the incident itself; 12 questions are about the effect of the incident; and one section is dedicated for additional comments. This information was provided through the auspices of the SAE Vehicle Sound for Pedestrian Committee. It is summarized here to describe incidents reported by a convenience sample, not a random sample. The information also shows the concerns expressed by this sample of 28 independent travelers who were involved in an incident with a vehicle. Twenty-five of the 28 people reported being blind or visually impaired. Seventeen use white canes, 7 use guide dogs, and 4 do not use mobility aids. Seven out of 28 reported to have moderate to mild hearing impairment. Individuals reported the location of the incidents as follows: driveway or alley (8); busy intersection (8); parking lot (6); side street (5); other (1). Most participants (20 out of 28) did not report the make, model, or propulsion system of the vehicle involved. Some of the listed models do not have a hybrid version. Individuals reported the background noise level at the time of the incidents as: relatively quiet (9); moderately noisy (7); somewhat noisy (6), very quiet (5); and very noisy (1). Five out of 28 pedestrians were struck or injured. The injuries include broken toe, contusions, back injuries, and concussion. Police reports were completed for two out of the 28 incidents. Vehicle turning, backing out, or moving into/from a driveway, alley, or parking lot were the most common concerns reported in the open-ended section of the survey.

Crash data on blind pedestrians and HEVs is limited. Five pedestrian injuries with quieter cars were reported by a convenience sample of 28 people. Common concerns included vehicles turning, backing out, or moving into/from a driveway, alley, or parking lots. Although the information from this questionnaire cannot be used to examine the factors contributing to pedestrian crashes the concerns are consistent with the results of the NCSA crash data analysis.

2.3 Blind Pedestrian Mobility Needs and Acoustic Cues for Travel

This section describes the information used by pedestrians to navigate, how the information is perceived by pedestrians who are blind, and how a reduction in auditory cues from vehicles may impact their decisions in various pedestrian-vehicle environments. This information is used to define the conditions under which the vehicle acoustic characteristics would be recorded and the detectability of vehicles evaluated (*critical safety scenarios*). The information was compiled based on the results of the following activities: (1) literature reviews; (2) observations of how blind pedestrians are trained to navigate in various pedestrian-vehicle environments; and (3) cognitive walkthroughs, where pedestrians who are blind and orientation and mobility specialists described cues and strategies at various pedestrian-vehicle environments.

2.3.1 Information and Strategies Used by Pedestrians Who Are Blind

Mobility depends in large part on perceiving the characteristics of the immediate surroundings. The information gathering and decision-making processes include several tasks, such as detecting a street and crossing location, identifying the type of traffic control device or traffic patterns, establishing a heading toward the opposite corner (alignment), determining a time to cross, and maintaining a straight path while crossing. People gather, interpret, and act on information about the environment by using multiple cues and more than one source of perceptual input. Pedestrians who are blind detect their arrival at an intersection using raised curb, slope of the curb ramp, detectable warnings, and traffic sounds among other wayfinding cues. Traffic sounds help them to orient themselves towards the crosswalks, to identify a time to cross, and to travel straight across the street (Blash, Wiener, & Welsh 1997); (Barlow, Bentzen, & Bond, 2005). The sound of traffic provides cues that help pedestrians identify vehicle operation (i.e., idling, accelerating, slowing) and vehicle maneuver (going straight, turning right or left). Vehicle operations provide information to assess the state of the traffic flow and to judge how much time they have to cross the street (NFB, 2008).

2.3.2 Conflicts at Sites Without Traffic Control

The risk at uncontrolled locations includes failure to detect approaching vehicles. Drivers failing to stop or yield to pedestrians are another risk. Previous studies have shown that drivers often do not yield to pedestrians, even those with mobility aids such as canes or guide dogs (Geruschat & Hassan 2005; Guth, Ashmead, Long, Wall, & Ponchillia 2005). Since pedestrians cannot rely solely on drivers to avoid a conflict they often need to identify a gap in traffic. The gap in traffic must be long enough to allow time to complete the crossing. Pedestrians who are blind rely on auditory cues to detect vehicles and identify gaps in traffic. The recommended crossing strategy at sites with low traffic volume is to cross when it is quiet. The technique is based on the premise that a vehicle will be loud enough to be heard far enough away to determine that it is safe to proceed when no masking sounds are present. The proliferation of quieter cars presents a risk at uncontrolled situations if these cars are not detectable with sufficient time before a pedestrian initiates the crossing.

Wall, Emerson, and Sauerburger (2008) have found that there are situations in which approaching traffic cannot be detected with sufficient time to avoid a conflict, regardless of vehicle type (e.g., HEV or ICE vehicle). They examined the factors that may affect the ability of pedestrians to detect oncoming traffic at crossing situations with no traffic control. The factors examined include: (1) level of ambient sound; (2) speed and sound emitted by approaching vehicles; and (3) physical factors (hills, roadway curvature, trees, and obstacles). The study identified situations where it is difficult for a pedestrian (with normal hearing and average walking speed) to detect approaching vehicles with sufficient time to determine that it is safe to proceed when no masking sounds are present. Twenty-three subjects (17 women and 6 men with vision impairments; mean age = 46) participated in the study conducted at three residential sites with mean ambient sound levels of 42 dB(A), ranging from 36 to 60 dB(A). Vehicle sound levels ranged from 56 dB(A) to 89 dB(A) while speed ranged from 13 mph to 60 mph. The performance measures included time-to-vehicle-arrival (the time from first detection of an approaching vehicle to the time when the vehicle passed in front of the pedestrian) and safety margin. Safety margin refers to the difference between the estimated crossing time and the time-to-vehicle-arrival. Estimated crossing time (7 seconds) assumes a walking speed 4 ft per second and roadway width of 28 ft, corresponding to the width of a two-way urban collector.

Ambient sound level was the strongest predictor of time-to-vehicle-arrival followed by physical factors (hills, roadway curvature, trees, and obstacles). As the ambient sound level increased, time-to-vehicle-arrival decreased. Ambient sound above 50 dB(A) (for the straight roadway scenario) and 38 dB(A) (for the hill scenario) negatively affected time-to-vehicle-arrival to unsafe levels. Overall, the ambient sound level, vehicle speed and sound, and physical factors, accounted for a third of the variability in time-to-vehicle-arrival. The study suggests how the strategy of crossing when quiet may be effective in some situations but not in others.

2.3.3 Conflicts at Stop-Controlled Intersections

Sauerburger (2005) describes the risk to blind pedestrians at two-way and four-way stop-controlled intersections. Four sources of risk are identified at two-way stop controlled intersections: (1) traffic turning into the crosswalk from the parallel street; (2) traffic approaching at the stop sign on the street the pedestrian is about to cross; (3) traffic waiting at the stop sign on the street the pedestrian is about to cross; and (4) traffic coming across the intersection from the other stop sign on the perpendicular street. The number of conflict points increase at four-way stop-controlled intersections. In particular, the risk includes stopped vehicles that surge forward and across from all directions and idling vehicles that are not audible.

Some of the strategies to reduce risk at sites with stop control as described by Sauerburger (2005) include:
1. Cross when there are no masking sounds and when no vehicles are heard.
2. Cross when traffic in the nearest parallel lane is approaching too fast to turn.

2.3.4 Conflicts at Signalized Intersections

Some of the strategies to reduce risk at signalized intersections as described by Sauerburger (2005) include:
1. Cross early in the cycle to reduce conflicts with vehicles turning left from the parallel lane. Cross when vehicles in the nearest half of the parallel lane are blocking traffic turning into the crosswalk.
2. Crossing early in the cycle when traffic is moving slowly and raising the hand (facing the driver) could also reduce conflicts with vehicles turning right from the parallel lane and into the crosswalk.
3. Crossing clockwise also reduces conflicts with left-turning vehicles. In this case, the pedestrian can clear the conflicting area sooner.
4. Cross clockwise to be more visible to the driver and reduce conflicts with vehicles turning right from the parallel lane.

2.3.5 Complex Non-controlled Intersections

There are more complex intersections where the surge of parallel traffic is intermittent or when the signal status cannot be determined by traffic sounds due to complex traffic signal timing, significant turning traffic, or when there is relatively high and continuous background noise. The perceptual problems faced by blind pedestrians at complex non-controlled locations such as roundabout and channelized turn lanes have been documented. A study by Guth, Ashmead, Long, Wall, and Ponchillia (2005) shows that blind pedestrians took significantly longer to report crossable gaps at single-lane roundabouts when compared to sighted pedestrians (on average 3 to 4 seconds more). This delay (i.e., shorter times-to-vehicle-arrival) may pose a safety risk to pedestrians who are blind since the approaching vehicle is now closer to the pedestrians at the time they decided to initiate the crossing.

Safety margins were computed based on the time from when the button is pressed (as an indication of detection of crossable gap), the remaining time until the next vehicle entered the crosswalk, and how long it would have taken to cross at a walking speed of 4 ft per second. Pedestrians' assessment was affected by the characteristics of the site, such as the geometry and traffic volume. For example, a low-volume, single-lane roundabout was about as safe for blind pedestrians as for sighted pedestrians since the gaps were sufficiently long enough that the increased detection latency for the blind was negligible. This study highlights the variability in pedestrian response due to traffic characteristics such as traffic volume, intersection geometry, and visual impairment. Pedestrians who are blind indicated that several factors affected their judgment of a crossable gap. These include the sound of traffic in the circulatory roadway, the masking sound of the vehicle as it passes the crosswalk, wet pavement, and wind (Guth et al., 2005).

Schroeder, Rouphail, and Wall Emerson (2006) examined crossing difficulties for pedestrians (9 blind and 9 sighted) at channelized turn lanes. Channelized lanes refer to the physical separation of conflicting traffic movements into distinct paths of travel. Travel paths are separated by a traffic island or pavement markings (AASHTO, 2004). Channelized turn lanes are particularly problematic for blind pedestrians because they are designed to permit continuous traffic flow. As with other uncontrolled sites, a pedestrian must detect a gap that is long enough to cross, detect vehicles in the turn lane, and identify whether a driver has yielded. The results show that pedestrians who are blind made more decisions, required more time to make a crossing decision, accepted gaps that were too short, and rejected more gaps than sighted pedestrians.

2.3.6 Parking Lots and Driveways

The NCSA analysis shows the incidence of pedestrian crashes for the combined maneuvers (slowing, backing out, entering or leaving a parking space or driveway) is higher for HEVs when compared to ICE vehicles. The potential sources of risk are vehicles moving at low speeds. Most pedestrians included in our study indicated that a vehicle backing out of a parking space is difficult to detect in part because it is difficult to predict when it will move. Some participants mentioned they tried to minimize potential conflicts by walking in the periphery of the parking lot and avoiding walking behind and between vehicles.

2.4 Vehicle Operation, Traffic Sounds, and Vehicle Detectability

Hearing sensitivity of an individual and the audibility of traffic sounds are important elements in orientation and mobility. The ability to use auditory cues determines how successful a vision-impaired pedestrian can become. Two types of errors are of primary concern for orientation and mobility: detection and localization. Detection error refers to a mistake in judging the presence of a relevant object or event. Localization error refers to a mistake in judging the direction of an object or event. Head movement is important in the localization of sounds. Localization errors become larger when a sound source moves more slowly (Ito, 1997). The likelihood of perceptual errors by pedestrians is influenced by several factors including absence or reduced information, lack of perceptual or motor skill, inattention, and willingness to take risks (Blash, Wiener, &

Welsh, 1997). This section describes previous and ongoing research on the effects of traffic sounds and potential safety risk for pedestrians.

An early study on the perception of distances (Coleman, 1963) shows that the loudness of a sound is a major factor in estimating the distance to its source. Other research suggests that absolute distance judgment is influenced by spectral balance (Butler, Levy, & Neff, 1980; Coleman, 1968). Barnecutt and Pfeffer (1998) found that judgment of relative distance of isolated sounds is complex and it is not based on loudness alone, in particular for traffic sounds.

One of the earliest studies on the use of auditory information for independent travel was conducted in the late 1990s. The goal was to estimate individuals' ability to make safe crossing decisions. Wiener and Lawson (1997) recorded the frequency and intensity of traffic sounds under different conditions and compared these to audiograms in an attempt to estimate the amount of traffic noise that is audible to a listener with normal hearing. An audiogram is a standard graph used to record hearing thresholds for sounds at different frequencies. Measurements were recorded for ICE-powered vehicles accelerating from a stop and approaching an intersection in a residential and a small-business intersection. Octave band levels were determined for sounds generated during the first 5 seconds of vehicle accelerations, which corresponds to an estimation of the time needed for a pedestrian to determine whether a vehicle is moving straight or is turning. Table 2 shows the sound intensity for each of the four scenarios. The results show that most traffic sounds can be found in the lowest frequency bands (< 250 Hz). However, audiometric thresholds in the 500 Hz to 4000 Hz range may be more important in assessing an individual's ability to detect traffic because people have greater sensitivity to sounds at those frequencies. Octave band levels were compared to minimum audible field values to estimate the amount of traffic noise that is audible to a pedestrian with normal hearing (audiometric hearing level).

Table 2. Sound Intensity by Vehicle Operating Condition (Wiener & Lawson, 1997)

ID	Operating Condition (Scenario)	Description	Number of Trials	Sound Intensity Range (dB)	Sound Intensity Mean(dB)
1	Accelerating from stop	Residential area, individual vehicles *accelerating from stop* at stop sign-controlled intersection, single-lane approach	40	77.8 to 90.4	86
2	Accelerating from stop	Small-business area, group of vehicles *accelerating from stop* at a traffic light (four lanes; one-way approach)	20	80.6 to 103.5	88
3	Accelerating from stop	Small-business area, group of *vehicles accelerating from stop* at a traffic light (two lanes; two-way approach)	20	88.8 to 102.4	94

ID	Operating Condition (Scenario)	Description	Number of Trials	Sound Intensity Range (dB)	Sound Intensity Mean(dB)
4	Approaching	Residential area, individual vehicles *approaching* a stop sign-controlled intersection, single-lane approach. Measured at 110 ft from the intersection, the vehicle was traveling perpendicular to the pedestrians' direction of travel, and there was no parallel traffic	24	70.0 to 83.1	74.4

Results suggest that an individual with moderate hearing loss should be able to detect vehicles in these scenarios because the threshold is lower than the estimated minimum hearing levels. Individuals with severe hearing loss may not be able to detect traffic accelerating from a stop in a residential area (scenario 1) or vehicle approaching (scenario 4). The authors acknowledged the value of frequency-specific thresholds. Lastly, the authors cited Whitener's (1981) predictions that EVs may be a greater danger for pedestrians than ICE vehicles due to a reduction in sound output.

2.4.1 Vehicle Approaching at Low Speed

Wiener, Naghshineh, Salisbury, and Rozema (2006) compared the sound intensity output of a traditional ICE vehicle (2004 Toyota Corolla) with that of a hybrid vehicle (2004 Toyota Prius) approaching at 30 mph. The purpose of the study was to determine whether there is a significant reduction in sound intensity from the HEVs that may impact pedestrians' ability to detect it when compared to the ICE vehicle. The study only gathered physical measurements. Data collection was completed on a private street close to a principal arterial. Microphone location simulated a pedestrian, 5 ft 6 in tall, standing perpendicular to traffic, 5 ft behind a crosswalk. Sound intensity for each vehicle was measured while the vehicle approached the pedestrian from a distance of 110 ft at 30 mph while coasting and also while powered by its engine. The purpose of the test was to measure the contribution of the engine versus tire and wind noise. The results in Table 3 show the difference in intensity (powered versus coasting) is small and similar for both vehicles.

Table 3. Sound Intensity of Vehicles Approaching at 30 mph (Wiener, Naghshineh, Salisbury, & Rozema (2006)

Description	Number of Trials	Vehicle Make/Model	Mean Intensity dB(A)
Vehicles approaching from 110 ft at 30 mph. Powered	5	Toyota/Corolla	70.1
	5	Toyota/Prius	72.4
Vehicles approaching from 110 ft at 30 mph. Coasting	5	Toyota/Corolla	69.7
		Toyota/Prius	72.0

Additional measurements were taken to determine at what distance (110, 198, 242, 286, and 330 ft) from the microphone the HEV could be identified. The Prius measured 10 dB above the ambient sound level only in the 2000-Hz-frequency band at a distance of 198 ft. A vehicle traveling at 30 mph requires approximately 200 ft to stop—including driver response time and distance traveled while braking.

A series of experiments conducted by Rosenblum (2008) suggests that an HEV traveling at 5 mph is harder to localize when compared to an ICE vehicle. Measures include listeners' ability to identify the direction of a vehicle approaching at 5 mph from a distance of 110 ft away and the response time for correct detection. Two vehicles were binaurally recorded approaching a listener at 5 mph. The purpose of binaural recordings is to reproduce the acoustic characteristics of the sound similar to how a human perceives it. Vehicles approached from the left or the right traversing 110 ft and passing 5 ft in front of the listener. Recordings were completed in a quiet parking lot. The recordings were then played to blindfolded listeners (college students) over headphones in a laboratory. Participants were asked to quickly and accurately identify from which direction (left or right) the vehicle was coming quickly and accurately. Trials were truncated at response.

The first experiment recorded a 2004 Toyota Prius and a 2005 Mustang (15 dB difference when passing). Results show high detection accuracy (subjects indicated correct direction of approaching vehicle) for both vehicles. Reaction time for correct response was different for both vehicles. The Mustang was detected 5.5 s (40 ft) before arrival and the Prius was detected 3.3 s (23 ft) before arrival (or 40 percent closer). A second experiment examined the 2006 Toyota Prius and a 2004 Honda Accord (13 dB difference when passing). Similarly, results show high accuracy for both vehicles. The Accord was detected 4.9 s (36 ft) before arrival and the Prius was detected 1.4 s (11 ft) before arrival (or 69 percent closer). There was a significant effect of vehicle type and direction interaction in both studies.

The experiments were repeated, this time adding the sound of two ICE vehicles idling, which increased the background noise by 8 dB. The third experiment used the same vehicles as the first experiment (a 2004 Toyota Prius and a 2005 Mustang). Results show high detection accuracy for both vehicles. The Mustang was detected 3.8 s (28 ft) before arrival and the Prius was detected 1 s (7 ft) before arrival (or 74 percent closer). There was a significant effect of vehicle and direction interaction in both studies. The fourth experiment used the same vehicles as the second experiment (a 2006 Toyota Prius and a 2004 Honda Accord). The Accord was detected 3.0 s (22 ft) before arrival while the Prius was detected 0.2 s (1.6 ft) after arrival. There was a significant effect of vehicle type.

Two additional experiments were conducted using blind listeners (mean age 46 and 43). This time the recordings were played back in a hotel, not in a laboratory. Results show similar trends, with blind listeners (older relative to previous group) requiring slightly more time to respond, and thus smaller safety margins. The 2005 Mustang was detected 4.3 s before arrival (versus 5.5 s). The Prius was detected 2.1 s (15 ft) before arrival (versus 3.3 s). Adding the sound of two ICE vehicles, the Mustang was localized 2.4 s (18 ft) before arrival (vs. 3.8 s). The Prius was detected 0.7 s (1.2 ft) after arrival (versus 1 s before). In both studies there was a significant effect of vehicle type with blind listeners.

In a separate study, the Japanese Automobile Standards Internationalization Centre (JASIC, 2009) compared the equivalent sound level (L_{Aeq}) between HEVs in electric mode and ICE vehicles. They also collected sound data for different levels of background noise (45.2 dB(A); 52.6 dB(A) and 61.7 dB(A)). JASIC combined vehicle sound and ambient sound and used these combined recordings to evaluate subjective perception of vehicle sounds. Twenty participants (vision status not reported) were instructed to press a button when they perceived a vehicle. The scenarios evaluated included: vehicle stationary and vehicle approaching at low speeds (6.5, 10, 15, and 20 km/h). Participants took longer to detect HEVs (in electric mode) than the ICE vehicles tested when the background noise level was low and the speed was about 9 mph or less. The study also suggested that the difference between the two vehicles that is associated with the background noise becomes smaller as the speed of the vehicle increases. The study suggested that the situations where it is necessary to improve the perception of HEVs (in electric mode) occur when moving below 20 km/h (approximately 12 mph).

2.4.2 Vehicle Stationary

JASIC (2009) evaluated the perception of sound for stationary vehicles with various levels of background noise. In the stationary scenario it is assumed that the pedestrian is standing 6.56 ft to the side and 6.56 ft in front of vehicle (measured from vehicle centerline). Twenty participants (vision status not reported) were asked to press a button when they perceived a vehicle. Participants did not detect the stationary HEV when there was background noise of 45.2 dB(A) and 52.6 dB(A). In contrast, approximately 95 to 100 percent of the subjects detected ICE vehicles.

2.4.3 Vehicles Accelerating from a Stop

Wiener, Naghshineh, Salisbury, and Rozema (2006) compared the sound intensity output of an ICE vehicle (2004 Toyota Corolla) with that of a hybrid vehicle (2004 Toyota Prius). Table 4 shows the acoustic output of each vehicle recorded while accelerating from a stop on a private street. This could be analogous to a vehicle accelerating from a stop at a stop-controlled intersection. The Corolla is 8 dB(A) louder than the Prius when operating with its ICE. The Prius accelerating in electric mode is 17 dB(A) quieter than the Corolla acceleration at a normal rate. Pedestrians with normal hearing should be able to detect hybrid vehicles. However, pedestrians with moderate hearing loss (threshold in the 41 to 55 dB range) might have problems hearing the hybrid vehicle in electric mode.

Table 4. Test Conditions in Wiener, Naghshineh, Salisbury, and Rozema (2006)

Description	Number of Trials	Vehicle Make/Model/ Power	Mean Intensity dB(A)
Vehicles accelerating from stop to a maximum of *18 mph* within 5 seconds.	5	Toyota/Corolla/ICE	70.0
	5	Toyota/Prius/ICE	62.2
Vehicles accelerating from stop to a maximum of *6 mph* within 5 seconds.	5	Toyota/Prius/Electric Motor	52.9

2.5 Driver Reaction Time, Braking Distance, and Pedestrian Safety

One approach to examining the safety risk associated with quieter cars is to compute the distance to the vehicle at the time it was detected (detection distance). The needed detection distance is computed from the vehicle speed and the pedestrian response time. This approach assumes that only pedestrians respond to a potential conflict. A second approach is to assume that the driver is the one who responds to a potential conflict. Sight distance is the length of the road ahead that is visible to the driver. This distance should allow a below-average driver, traveling at or near the design speed, to stop before reaching a stationary object in his/her path. Sight distance has two components: brake reaction and braking distance (AASHTO, 2004). Brake reaction is the distance traveled from the time the driver detects an object to the instant the driver applies the brakes. The recommended design criterion for brake reaction time is 2.5 seconds. A 2.5-second brake reaction time for stopping situations considers the capabilities of most drivers, including older drivers. Braking distance refers to the distance needed to stop the vehicle once the drivers applied the brakes, and depends on vehicle speed, deceleration rate, and roadway grade (AASHTO, 2004). For example, the sight distance for a vehicle approaching at a constant 6 mph is 25.5 ft (assuming brake reaction time of 2.5 s and a constant deceleration rate of 11.2 ft/s^2). The vehicle would travel approximately 22.5 ft while the driver reacts. The driver would need another 3.5 ft to stop the vehicle. Drivers' sight distance can be compared against the pedestrian detection distance as a measure of risk. In this example, the pedestrian must detect the vehicle (and respond) when the vehicle is at least 25.5 ft away in order to avoid a potential collision.

3. CRITICAL SAFETY SCENARIOS FOR PEDESTRIANS WHO ARE BLIND

A series of critical safety scenarios were defined and discussed with pedestrians who are blind and with orientation and mobility specialists. Scenarios were defined by combining pedestrian-vehicle environments, vehicle type, vehicle maneuver/speed/operation, and ambient sound level. The risks at various pedestrian-vehicle environments were considered. These include vehicles approaching at a constant speed, vehicles turning into the pedestrian's path, and vehicles backing out into the pedestrian's path. In addition to these risks, pedestrians who are blind identified the kinds of information that helps them to navigate. This includes: vehicle presence; vehicle position relative to the pedestrian; vehicle direction of travel; and vehicle rate of acceleration or speed. Information is used to judge how fast the vehicle is moving or how soon the vehicle may reach the pedestrian position or travel path. Critical safety scenarios are described in this section.

3.1 Vehicle Backing Out

Some HEVs can use the electric motor as the sole source of propulsion for low speed and low acceleration driving, such as backing out; this is particularly true when the batteries are charged. There is a concern these vehicles may not be detectable when backing out. This task is complex for pedestrians since it is difficult to anticipate where there may be a driveway and when a vehicle will move out of a driveway. In addition, a driver's visibility may be limited. The pedestrian may have very limited time to respond to avoid a conflict. Thus it is important to include as a critical scenario vehicles backing out at low speed (as if they are coming out of a driveway). This scenario was included in the human subjects study (see Chapter 5.2.1). The traveling situation includes a pedestrian walking along a sidewalk with driveways on the left side; the pedestrian will hear distant vehicles in the background in all trials. This is similar to walking in an area that is a few blocks away from a main road. In some trials there will be the sound of a nearby vehicle backing towards the pedestrian at a constant speed of 5 mph.

3.2 Vehicle Traveling in Parallel and Slowing

Pedestrians who are blind often need to distinguish between a vehicle moving through an intersection and a vehicle turning into their path. The pedestrian needs to perceive this information when the vehicle is in the parallel street, before it turns into their path. The sound of slowing vehicles in the parallel street helps pedestrians identify turning vehicles. A quieter car slowing may not be detectable. This study includes data for vehicles slowing, from 20 to 10 mph, as if the vehicle is preparing to turn right from the parallel street. This scenario was also included in the human subjects study (see Chapter 5.2.2). The traveling situation includes a pedestrian trying to decide when to start across a street with the signal in his/her favor and there is a surge of parallel traffic on the immediate left. In some trials, a vehicle will continue straight through the intersection at 20 mph, so the pedestrian can cross whenever they choose. However, in some trials there will be a vehicle slowing down from 20 mph as if to turn right into the pedestrian path. The pedestrian must be able to detect when a vehicle is slowing.

3.3 Vehicle Approaching at Low Speed

One of the strategies used by pedestrians who are blind is to cross when the road is quiet. The technique assumes that a vehicle is loud enough to be heard far enough away to determine that it is safe to proceed when no masking sounds are present and no other vehicles are detected. Preliminary studies have shown that HEVs approaching at low speed (less than 12 mph) may not be detectable. A quieter car approaching at low speed may not be detected until it is too close to the pedestrian. This scenario was included in the human subjects study for vehicles approaching at 6 mph (see Chapter 5.2.3). The traveling situation includes a pedestrian standing on the curb waiting to cross a one-way street where there may be vehicles approaching from the left. There are vehicles in the background in all trials. The pedestrian must be able to detect a vehicle that would affect the decision about when to start across the street.

The difference in sound levels between HEVs and ICE vehicles may become smaller as the speed of the vehicle increases. Both the electric motor and engine are used to propel HEVs at higher speeds. In addition, other cues such as tire noise are more noticeable at higher speeds. The study includes acoustic data for vehicles approaching at low (6 mph, 10 mph) and moderate speeds (20 mph, 30 mph, 40 mph) to examine how the acoustic characteristics of HEVs and ICE vehicles differ as a function of vehicle speed.

3.4 Vehicle Accelerating from a Stop

Pedestrians who are blind use the sound of traffic in the parallel street to establish alignment and to identify a time to cross. The sound of accelerating vehicles in the parallel street indicates, for example, that the perpendicular traffic does not have the right of way and thus a crossing opportunity is available. A safety concern is that quieter cars may not be heard during initial acceleration. Pedestrians may initiate their crossing as soon as they detect the surge of parallel traffic or may delay the decision to make sure traffic is moving straight through the intersection and not turning into their path. A significant delay in detecting the surge of parallel traffic may impact the blind pedestrian's ability to complete a crossing within the designated walking interval. This study records data for vehicles accelerating from a stop and examines how the acoustic characteristics of HEVs and ICE vehicles differ during initial acceleration.

3.5 Vehicle Stationary

Finally, there is a concern that a quieter car may not be detected when it is stationary and idling. The sound of vehicles idling provides important cues. For example, in the far lane it gives cues about the width of the road (number of lanes), conveying information about the distance to walk, and the time required to cross the road. A quieter car may not be detected when it is stationary at intersections or parking lots and it may start moving suddenly at the same time the pedestrian enters the conflicting path. Previous studies suggest that a stationary HEV is not detectable even when the background noise is moderate (JASIC, 2009). The current study measured acoustic characteristics for vehicles in this operating condition.

3.6 Ambient Sound Levels

The intention of the scenarios to be evaluated is to model situations where pedestrians could expect to detect vehicles using auditory cues. A few sites were identified by pedestrians who are blind and travel by themselves on a regular basis. In particular, discussion led to the conclusion that ambient sound levels used for the human subject studies should simulate a quiet suburban neighborhood and a somewhat noisier suburban neighborhood. Simulation of a noisier urban environment was excluded because individuals who are blind indicated that they would often avoid making such crossings without assistance. In addition, preliminary data and previous studies suggest that the difference in detectability between HEVs and ICE vehicles is small when ambient sound levels are too high, with pedestrians having difficulty detecting either type of vehicles.

4. ACOUSTIC MEASUREMENT OF VEHICLES AND AMBIENT SOUNDS

4.1 Procedure to Measure Vehicles and Ambient Sounds

Sounds emitted by HEVs and ICE vehicles operated in various conditions were measured and recorded. Acoustic measurements were recorded for the following operating conditions: (1) vehicle backing out at 5 mph (mimicking a vehicle backing out of a driveway); (2) vehicle slowing from 20 to 10 mph (mimicking a vehicle preparing to turn right from the parallel street); (3) vehicle approaching at a low constant speed (6 mph and 10 mph); (4) vehicle accelerating from a stop; and (5) vehicle stationary. These conditions simulate vehicle operation in critical safety scenarios. Additional measurements were collected for vehicles approaching at moderate constant speeds (20 mph, 30 mph, and 40 mph). Ambient sounds were recorded and combined with vehicle audio recordings for use in the human subject studies.

4.1.1 Objectives

The purpose of the acoustic measurements is to: (1) collect objective data to characterize the sound emitted by both HEVs and ICE in various operating conditions (simulating vehicle operations in critical safety scenarios); (2) acquire binaural audio recordings of these vehicles; and (3) acquire binaural audio recordings of ambient sound levels that are representative of locations where pedestrians could expect to hear a nearby vehicle. The information is used to document the overall sound levels and spectrum for HEVs and ICE vehicles. It is also used to examine the overall sound levels for the two types of vehicles as a function of vehicle speed. This section of the study describes the vehicles used, measurement sites, instrumentation and equipment layouts, data collection protocols, and results.

4.1.2 Test Vehicles

Test vehicles included three HEVs (Honda Civic, Toyota Prius, and Toyota Highlander) and their ICE twins (the Toyota Matrix serves as a twin for the Toyota Prius). Vehicle make, model and engine type are listed in Table 5. Vehicle model years ranged from 2008 to 2010. Five of the six vehicles were new 2009-2010 vehicles. Vehicles were in good operating condition and did not generate sounds from a defect in the conditions of the vehicle. Tires had a tread depth considered sufficient for safe operation, even or no wear, and were representative of standard OEM tires. Tires were inflated to the manufacturers' recommended pressure and debris in the treads was removed prior to testing (Figure 1).

Table 5. Test Vehicles, Make, Model Year and Engine Type

Vehicle Make	Vehicle Model/Year	Engine Type
Toyota	Prius (2010)	Hybrid
Toyota	Matrix(2009)	Combustion

Vehicle Make	Vehicle Model/Year	Engine Type
Honda	Civic (2009)	Hybrid
Honda	Civic (2009)	Combustion
Toyota	Highlander (2009)	Hybrid
Toyota	Highlander (2008)	Combustion

Figure 1. Tire Tread Showing Debris Removed Prior to Testing

4.1.3 Vehicle Measurement Site

The background noise at the time of measurement affects the ability to quantify the sound levels of quieter vehicles. Significant efforts were made to locate a vehicle measurement site with very low background noise in order to measure quieter vehicles in critical safety scenarios. Some HEVs were not measurable under some operating conditions. This is due to the extremely low sound level of vehicles operated in electric-only mode relative to the existing background. Such cases were noted during the measurements and analysis.

Site requirements included the following:
1. Minimal background noise sources, e.g., birds, insects, wind-induced noise, and other distant vehicles;
2. Low wind speeds, e.g., less than 10 mph;

3. Closed, flat course with sufficient distance to bring the vehicle to the beginning state of the test condition outside the audible range;
4. Clean asphalt pavement in good condition;
5. Flat open area adjacent to the vehicle travel path for instrument setup; and
6. No other vehicles on nearby tracks.

The vehicle measurement site is located at NHTSA's Vehicle Research and Test Center, at the Transportation Research Center in East Liberty, Ohio. Most of the data was recorded during the night (with no other vehicle operating at the test facility) to guarantee the quietest possible background levels at the site and obtain the highest quality recordings possible. An aerial photo of the site is shown in Figure 2. Sub-site 1 was used for idle and acceleration measurements while sub-site 2 was used for all other measurements. Figure 3 and Figure 4 illustrate sub-site 1.

Figure 2. Aerial View of Vehicle Measurement Site

Figure 3. Vehicle Measurement Sub-Site 1 Facing Southeast

Figure 4. Vehicle Measurement Sub-Site 1 Facing Northeast

4.1.4 Instrumentation and Equipment Layout

Two sub-sites were used at TRC. Sub-site 1 was used for measuring idle and acceleration measurements. Sub-site 2 was used for all other measurements (i.e., those requiring more than 200 ft of flat roadway. Note: The roadway sub-site was found to have a number of small bumps and cracks that limited its usable length to approximately 200 ft; hence the use of sub-site 2 for tests requiring more space). The equipment layout is shown in Figure 5 for sub-site 1 and in Figure 6 for Sub-site 2. The microphone setup captured binaural audio recordings and sound level measurements at a location representative of a pedestrian attempting to accomplish tasks as described in the critical safety scenarios in Chapter 3. Acoustic data was collected 12 ft from the center of the travel lane 5 ft above the ground using a sound level meter (SLM) (Larson Davis Model 824 (LD824)) real-time spectral analyzer. A second sound level meter was set up 50 ft from the center of the vehicle travel lane 5 ft above the ground to provide additional acoustic data that could be used in further analyses. The other two microphones were part of the binaural head system (Neumann KU 100 Binaural Stereo Microphone system). The microphones of the binaural head were placed in position corresponding to the anatomical location of the human ears. Data from the sound level meter was stored on their internal memory and audio recordings. Binaural head data were stored on secure digital cards as 16 bit .wav files using two digital audio recorders (Sound Devices 744T). The time that a vehicle passed the binaural head was recorded on a third channel of the audio recording using a triggered tone generator. The control of the all microphones was centralized at the instrument table situated 100 ft from the 12 ft microphones. Temperature, relative humidity, and wind speed were monitored using a meteorological system (Qualimetrics TAMS).

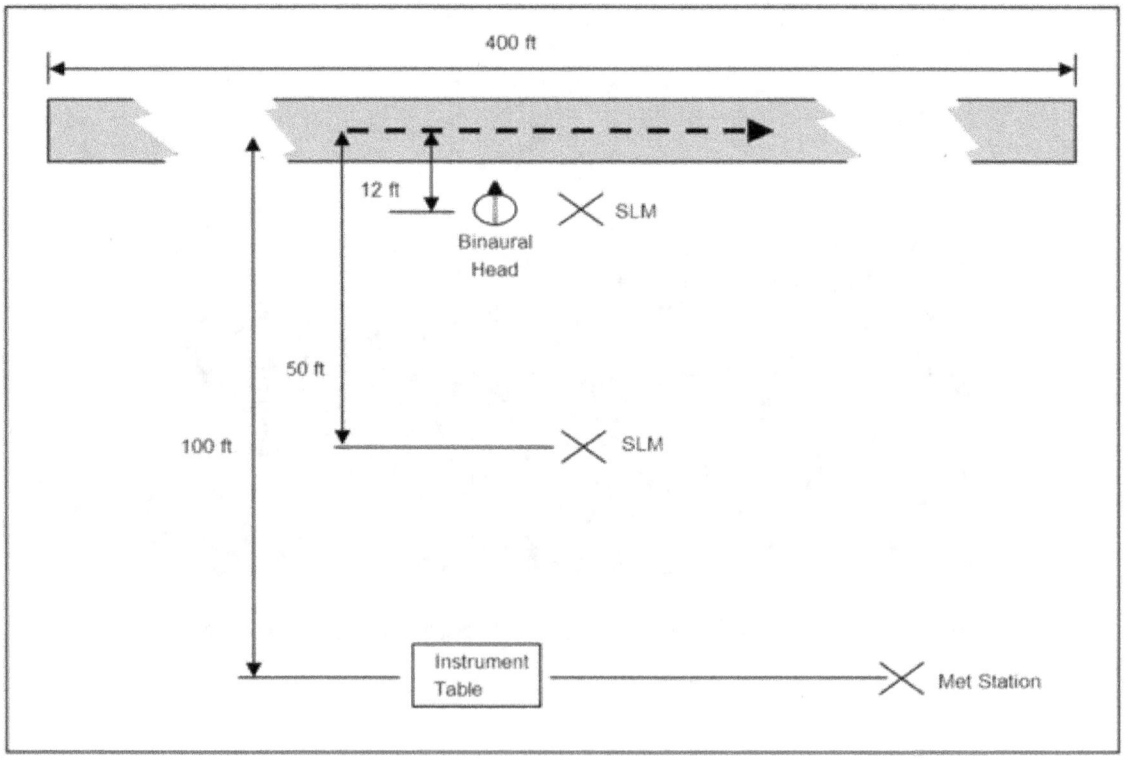

Figure 5. Vehicle Measurement Equipment Layout, Sub-Site 1

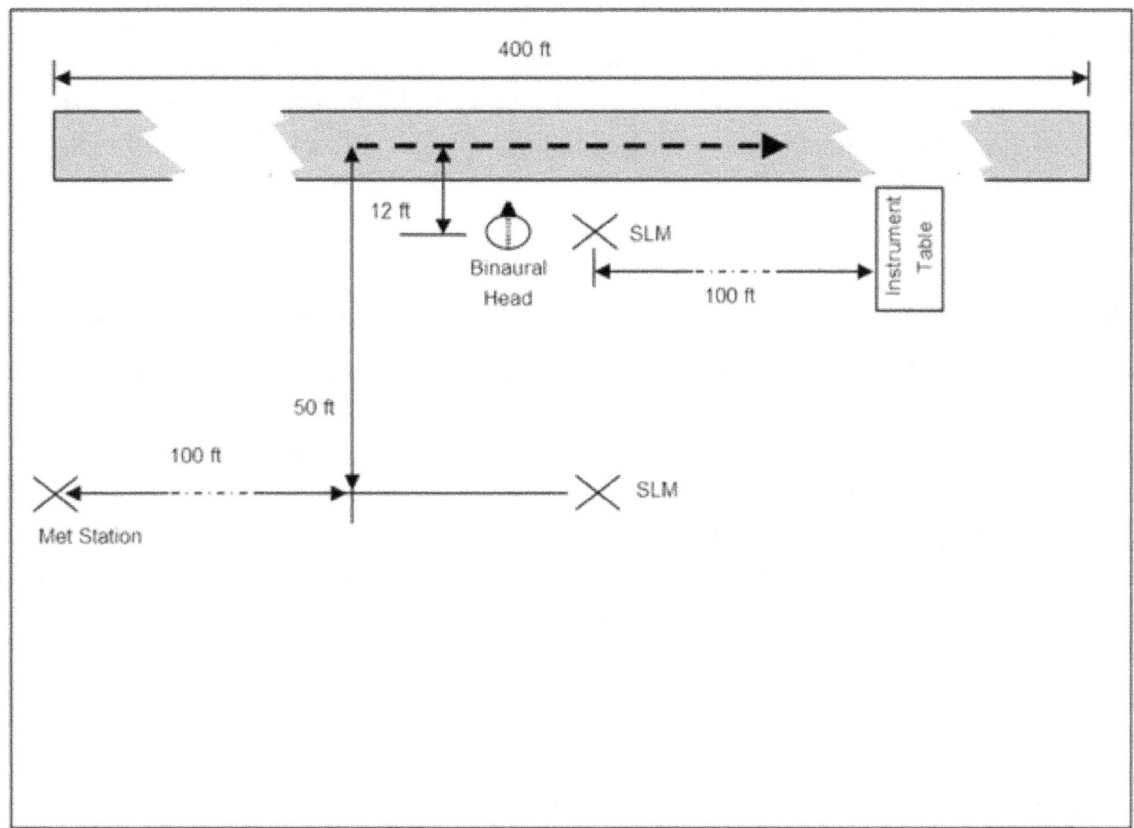

Figure 6. Vehicle Measurement Equipment Layout, Sub-Site 2

4.1.5 Data Collection Protocol

A data collection procedure was required to ensure vehicle acoustic parameters were measured consistently. SAE has developed a draft test procedure "Measurement of Minimum Noise Emitted by Road Vehicles" (SAE J2889-1) to measure the acoustic characteristics of vehicles at low speed. The procedure was reviewed to assess its suitability to collect vehicle acoustic data in critical safety scenarios. A modified procedure was developed for this Quieter Cars study because the SAE J2889-1 draft test procedure currently does not include provisions to collect data that can be used to test pedestrians' responses to vehicle acoustic parameters in critical safety scenarios. The measurement procedure in this study follows recommendations of the SAE draft procedure recommendation with regard to instrument settings; however, the study deviates from the SAE procedure with respect to operating condition, height, distance, and orientation of the microphones. The reason is that the goal of this study is to document vehicle acoustics under "critical safety scenarios" rather than the "minimum noise emitted." For example, data was recorded using a binaural head in addition to sound level meters.

Data recorded include the minimum A-weighted level (L_{Amin}), average A-weighted level ($L_{Aeq0.5s}$), maximum A-weighted level (L_{Amax}), and average unweighted one-half-second one-third octave band sound levels for the following vehicle operating conditions:

1. Vehicle backing out at 5 mph;
2. Vehicle slowing from 20 mph to 10 mph;
3. Vehicle approaching at low speeds (6 mph and 10 mph constant speed pass);
4. Vehicle approaching at moderate to high speeds (20 mph, 30 mph, 40 mph constant speed pass);
5. Vehicle accelerating from a stop; and
6. Stationary vehicle.

The following site conditions were required for a measurement to be considered acceptable:

1. No significant meteorological or environmental effects, e.g., wind speeds < 10 mph;
2. Ambient levels sufficiently low such that maximum sound pressure level is at least 3 dB greater than the ambient; and
3. Minimal extraneous sounds such as bird or animal sounds.

A correction was applied during analysis of the sound meter data to adjust for contamination due to the background when the maximum sound pressure level was less than 15 dB above the background level:

$$SPL_{corrected} = 10\log_{10}\left(10^{(SPL_{measured}/10)} - 10^{background/10}\right)$$

The measurement system setup consists of the following steps:

1. Clocks of all measurement equipment were synchronized by using a GPS-received Universal Time Code (UTC).
2. Preliminary sound level calibrations of the acoustic systems were performed to ensure that all equipment was operating properly.
3. The electronic noise floor of each system was determined by using a non-transducive (i.e., mechanically passive) capacitive load and measuring and storing 30 seconds of data.
4. After reinstallation of the microphone, a pre-measurement sound level calibration was performed.
5. A windscreen for each microphone was deployed and the preamplifier cable was secured to the mast and/or the leg of the tripod to prevent vibration and audible interference. For the sound level meters, the measurement mast was oriented vertically.
6. Continuous meteorological data was collected.

At the start of each event, all acoustic systems (SLMs and binaural head) were started. The systems were then monitored by two technicians, who performed the following tasks:

1. Noted the maximum sound level (L_{Amax}) observed on the LD824 on the log sheet;
2. Noted any sounds heard from the measured vehicle, such as electric motor, engine, compressor, fan, tire noise, etc., on the log sheet;

3. Noted whether or not the recording instrumentation indicated a minimum 3-dB rise and fall;
4. Noted any potentially contaminating sounds that were heard;
5. If possible, observed that wind speeds did not exceed the predetermined limit of 10 mph;
6. Reset the "current" memory of the LD824s in preparation for the next event; and
7. Prepared the log sheets for the next event.

The following tasks were completed every four hours and at the end of the measurement day:

1. Checked instrument clock synchronization, if necessary, re-synchronized; any differences were noted on the log sheet.
2. The system was powered down and disassembled at the end of each measurement day.
3. The gel cells were setup for overnight charging.
4. The data from the SLMs were downloaded and briefly checked to ensure the systems were functioning properly and all events were captured.

4.1.6 Data Collection Procedure for Each Vehicle Operation

4.1.6.1 Constant Speed Passby Measurement Procedure

1. Vehicle accelerates to a constant specified speed:
    a. Target speeds include 5 (reverse only), 6, 10, 20, 30, or 40 mph
2. Target speed attained outside audible zone
3. Target speed maintained within a tolerance of +/- 1 mph
4. Target speed maintained for at least 100 ft beyond the microphone line
5. For HEVs all practicable attempts made to maintain electric motor only vehicle:
    a. Propulsion mode documented in the measurement log for each passby. Options include ICE only, Electric Motor only, and ICE and Electric Motor mix.
6. For both HEVs and ICE vehicles all practicable attempts made to operate vehicle with accessory devices (such as cooling fans) off
7. A minimum of four repetitions for each operating condition measured for the purpose of obtaining at least one clean measurement, i.e. one suitable for human subject testing

4.1.6.2 Accelerating Passby Measurement Procedure

1. Vehicle starts at rest at a distance of 200 ft from the binaural head (as measured along the road) and then accelerates at a constant rate to a speed of 20 mph.

The driver attempts to accelerate at the same rate for each vehicle for each repetition
2. For both HEVs and ICE vehicles all practicable attempts made to operate vehicle with accessory devices (such as cooling fans) off
3. A minimum of four repetitions measured

4.1.6.3 Decelerating Passby Measurement Procedure

1. Vehicle accelerates to a constant specified speed of 20 mph outside of the audible zone.
2. Driver brakes 100 ft from the microphone line at a constant rate of 1 m/s² and reaches 10 mph at the microphone line

    Where,

    $$t = (v_1 - v_0)/a_{const}, \quad x(t) = \frac{1}{2}a_{const}t^2 + v_0 t$$

3. After the measurement, driver relays the final speed that was achieved as the vehicle passed the binaural head to the technicians maintaining the logs
4. For both HEVs and ICE vehicles all practicable attempts made to operate vehicle with accessory devices (such as cooling fans) off
5. A minimum of four repetitions measured

4.1.6.4 Stationary Measurement Procedure

1. Vehicle starts and remains at rest adjacent to the binaural head
2. For HEVs power is on, but engine and all accessory devices, e.g., compressors, radios, and cooling fans, are off
3. For ICE vehicles engine running at idle with the vehicle in park, but all accessory devices, e.g., compressors, radios, and cooling fans, are off
4. Measure for a minimum of 1 minute

4.2 Results of Acoustic Measurement of Vehicle and Ambient Sounds

4.2.1 Vehicle Acoustic Measurement Results

Objective metrics characterize the sound emitted by the vehicles measured under the specified operating conditions. Two of the three HEVs (Toyotas) were recorded while operating in electric mode for all runs except for those greater than 20 mph. Results for the 12-ft microphone location are discussed for the three operations used in the human subject studies (see Chapter 5.2) and for stationary vehicles. Additional results are listed in Appendix A.

4.2.1.1 Corrections for the Background Noise

The background noise at the time of measurement affects the ability to quantify the sound levels of vehicles. Sound level measurements for a vehicle cannot be reported accurately when the sound level produced by the vehicle is too close to the background noise level (i.e., less than 3dB). The tabulated results for those events are labeled as "background" to indicate that they were not sufficiently above the background level. A correction to the sound level reading is needed when the difference between the sound pressure level for the ambient and the vehicle is between 3 to 15 dB.

Time histories for the sound levels were analyzed to determine if a correction for the background noise was needed. Three cases were considered: (1) the sound levels at the microphone were sufficiently higher than the background to be used without correction; (2) the sound levels at the microphone were sufficiently higher than the background to be used with correction; and (3) the levels at the microphone were too low for the event to be used. Figure 7 shows an example of a passby event that does not require correction for the background level. This is easily observable because the passby, which occurs at time equal to 10 seconds, is more than 15 dB above the background level, as indicated by the second dotted black line. In Figure 8, a sample is shown of a passby event that requires a correction to the level at passby. Here it can be seen that the passby level falls within the range of 3 dB to 15 dB greater than the background level, as specified by SAE J2889-1. In this case the reported levels at passby are corrected according to:

$$SPL_{corrected} = 10\log_{10}\left(10^{(SPL_{measured}/10)} - 10^{background/10}\right)$$

In Figure 9, a sample is shown where the event is not sufficiently above the background (falling below the background + 3dB line) and therefore is considered too low for correction. In such a case the tabulated results for the event are labeled as "background" to indicate that they were not sufficiently above the background level.

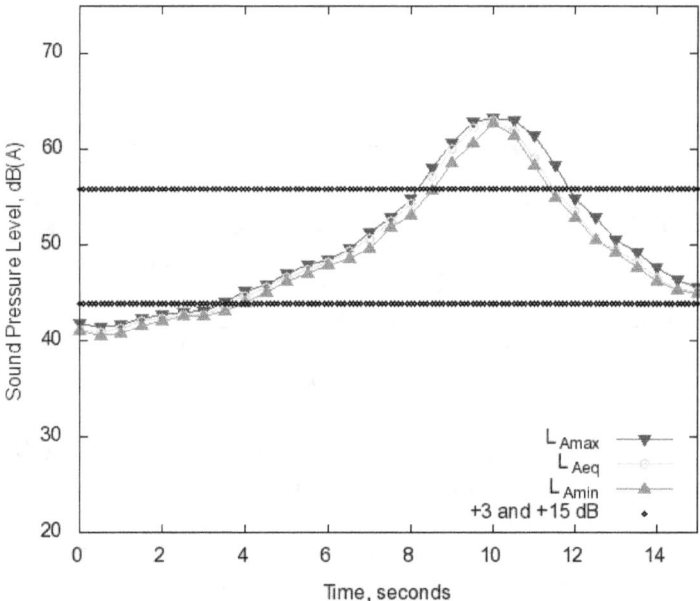

Figure 7. Sound Pressure Level in dB(A) for Prius 20 mph Passby (No Correction Needed)

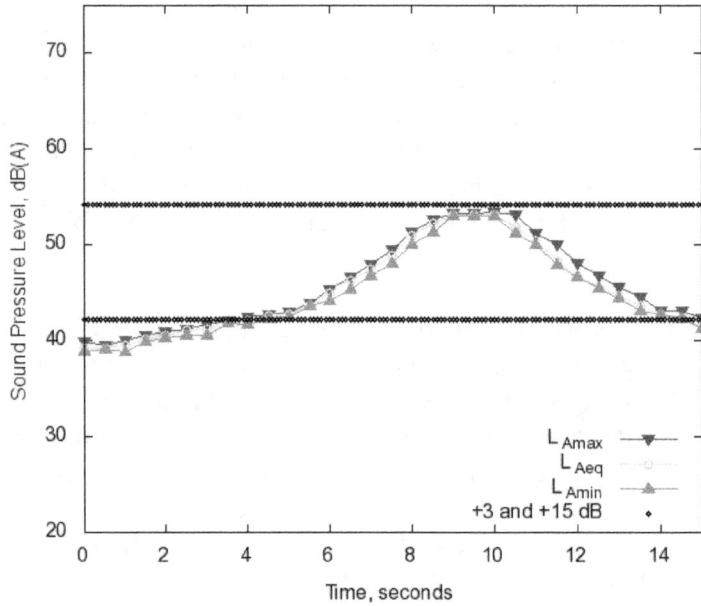

Figure 8. Sound Pressure Level in dB(A) for Prius 10 mph Passby (Suitable for Correction)

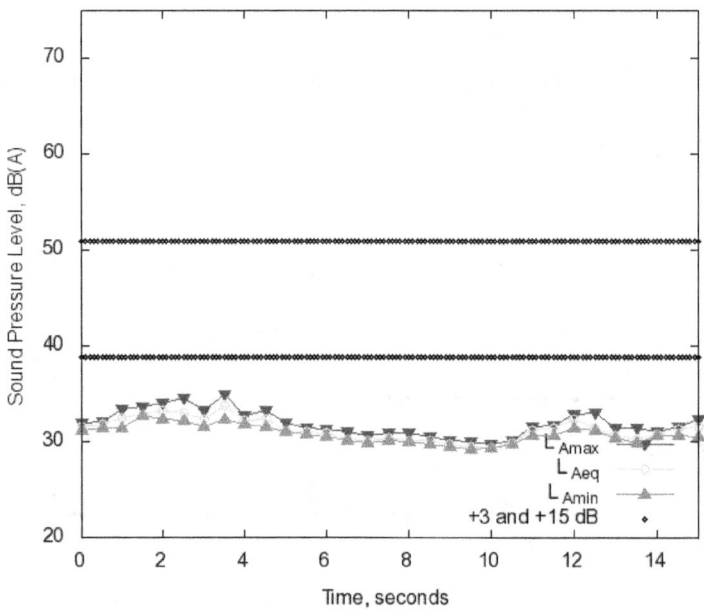

Figure 9. Sound Pressure Level in dB(A) for Prius Idle (Too Close to Background for Correction)

A complete set of sound pressure level time histories is contained in Appendix A.1.

4.2.1.2 Broadband Metrics at 12-ft Microphone Location

Broadband metrics, minimum A-weighted level (L_{Amin}), average A-weighted level ($L_{Aeq0.5s}$), and maximum A-weighted level (L_{Amax}), are given for the three operations used in the human subject studies (Tables 6 to 8) and for stationary vehicles (Table 9). These measurements are taken at the 12-ft microphone location. In general, the HEVs tested had lower overall sound levels for each of the conditions used in the human subject studies at the 12-ft microphone location. Table 6 shows the passby sound levels for 5 mph reverse constant speed passby events. The overall sound levels for HEVs operated in reverse at 5 mph are 7 to 10 dB(A) lower than the sound level for their ICE vehicle twins. The overall sound levels for HEVs do not differ considerably from the sound levels for their ICE twins in the deceleration maneuver as shown in Table 7 . Table 8 shows the sound levels for vehicles approaching at a constant speed of 6 mph. The overall sound levels for HEVs are 2 to 8 dB(A) lower than the sound levels for their ICE vehicle twins operated at 6 mph.

Table 6. Passby Levels for Reverse Constant Speed Passby Events at the 12-ft Microphone Location

Vehicle Type	Operation	L_{Amin}	L_{Aeq}	L_{Amax}
Prius	Reverse	43.7	44.2	44.8
Matrix	Reverse	51.2	51.3	51.5
Civic Hybrid	Reverse	48.5	48.5	49.0
Civic	Reverse	58.0	58.2	58.9
Highlander Hybrid	Reverse	44.6	45.9	48.6
Highlander	Reverse	52.3	52.7	53.1

Table 7. Passby Levels for Deceleration Passby Events at the 12-ft Microphone Location

Vehicle Type	Operation	L_{Amin}	L_{Aeq}	L_{Amax}
Prius	Deceleration	52.2	53.0	53.4
Matrix	Deceleration	53.8	54.2	54.6
Civic Hybrid	Deceleration	55.7	56.6	57.2
Civic	Deceleration	54.8	55.0	55.3
Highlander Hybrid	Deceleration	52.2	53.0	53.7
Highlander	Deceleration	54.9	55.4	55.8

Table 8. Passby Levels for 6 mph Constant Speed Passby Events at the 12-ft Microphone Location

Vehicle Type	Operation	L_{Amin}	L_{Aeq}	L_{Amax}
Prius	6 mph	44.4	44.7	45.1
Matrix	6 mph	53.0	53.5	54.2
Civic Hybrid	6 mph	49.2	49.3	49.5
Civic	6 mph	51.8	52.0	52.6
Highlander Hybrid	6 mph	52.5	53.2	54.9
Highlander	6 mph	55.2	55.5	55.9

Table 9 shows the sound levels for idle. Overall sound pressure levels for the Toyota hybrids, tested when stationary, were too low to be recorded for the background condition present. The overall sound level for ICE vehicles when idling range from 46 to 48.1 dB(A). Results for 10 mph, 20 mph, 30 mph, 40 mph, and acceleration are listed in Appendix A.2.

Table 9. Overall Levels by Vehicle for Idle at the 12-ft Microphone Location

Vehicle Type	Operation	L_{Amin}	L_{Aeq}	L_{Amax}
Prius	Idle	Background	Background	Background
Matrix	Idle	47.6	47.8	48.1
Civic Hybrid	Idle	44.6	44.8	45.1
Civic	Idle	45.8	46.0	46.4
Highlander Hybrid	Idle	Background	Background	Background
Highlander	Idle	47.9	48.1	48.5

4.2.1.3 Spectral Shape at the 12-ft Microphone Location

Sounds at the same level may be more or less detectable depending on their spectral shape. A sample one-third octave band spectrum is shown in Figure 10 for vehicles traveling in reverse at 5 mph. Figure 11 shows the one-third octave band spectrum for vehicles decelerating. Lastly, Figure 12 shows the one-third octave band spectrum for vehicles approaching at a constant speed of 6 mph. These spectra, while not showing the finest spectral detail, provide a means of making general comparisons of spectral balance of the various measured events. There is a slight trend for HEVs (at low speeds) to have less high frequency content relative to the overall sound level. An exception is a peak in the Toyota vehicles in the 5000 Hz one-third octave band, especially while decelerating/braking. A more detailed narrow-band spectral analysis of the sound emitted by a Prius tested during the slowing maneuver is presented in Section 5.2.2. Results for 10 mph, 20 mph, 30 mph, 40 mph, and acceleration are listed in Appendix A.3.

Chapter 4: Acoustic Measurement of Vehicles and Ambient Sounds

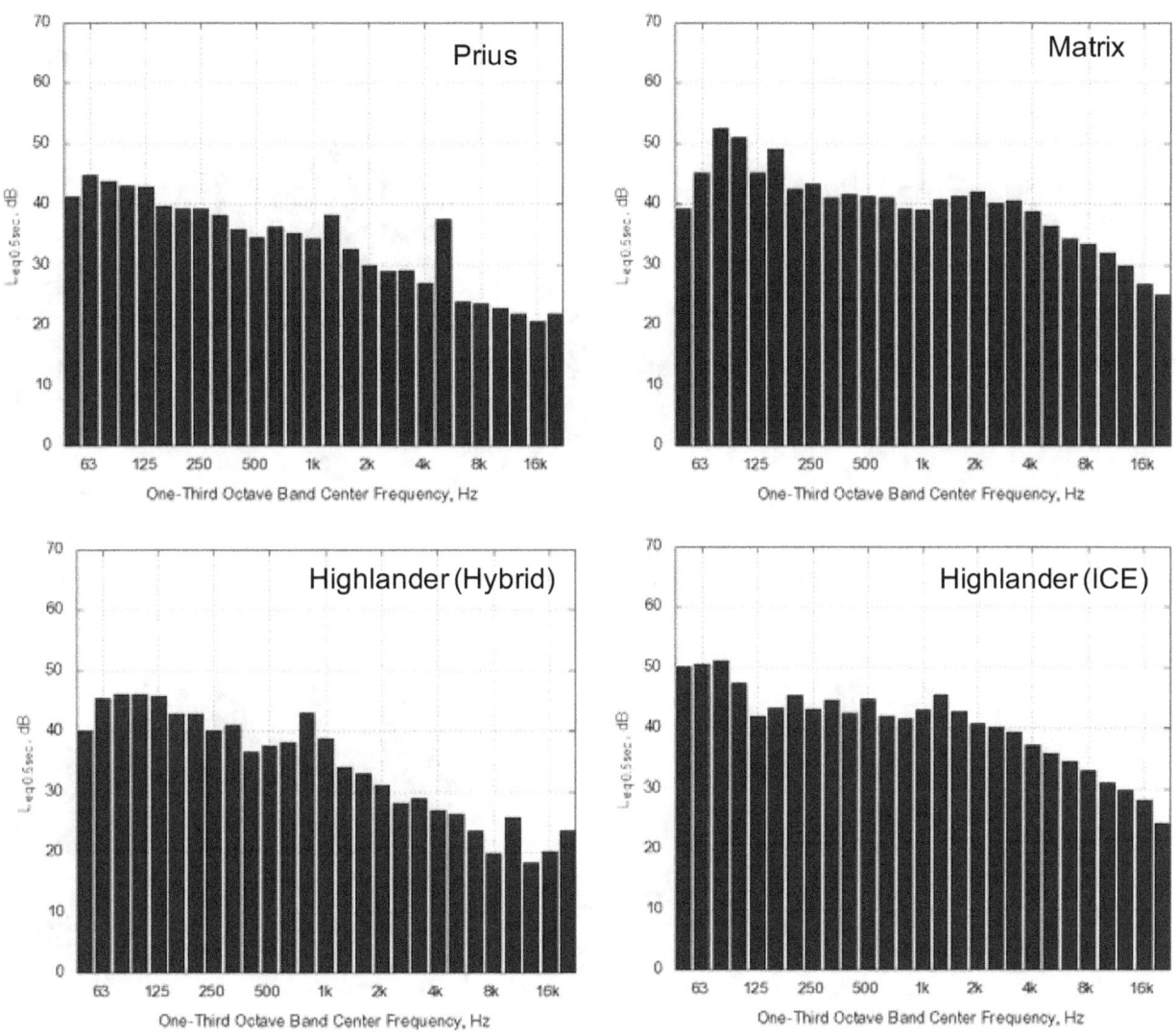

Figure 10. Sample One-Third Octave Band Spectra for Reverse 5 mph Constant Speed Passby at 12 ft Microphone Location

Chapter 4: Acoustic Measurement of Vehicles and Ambient Sounds

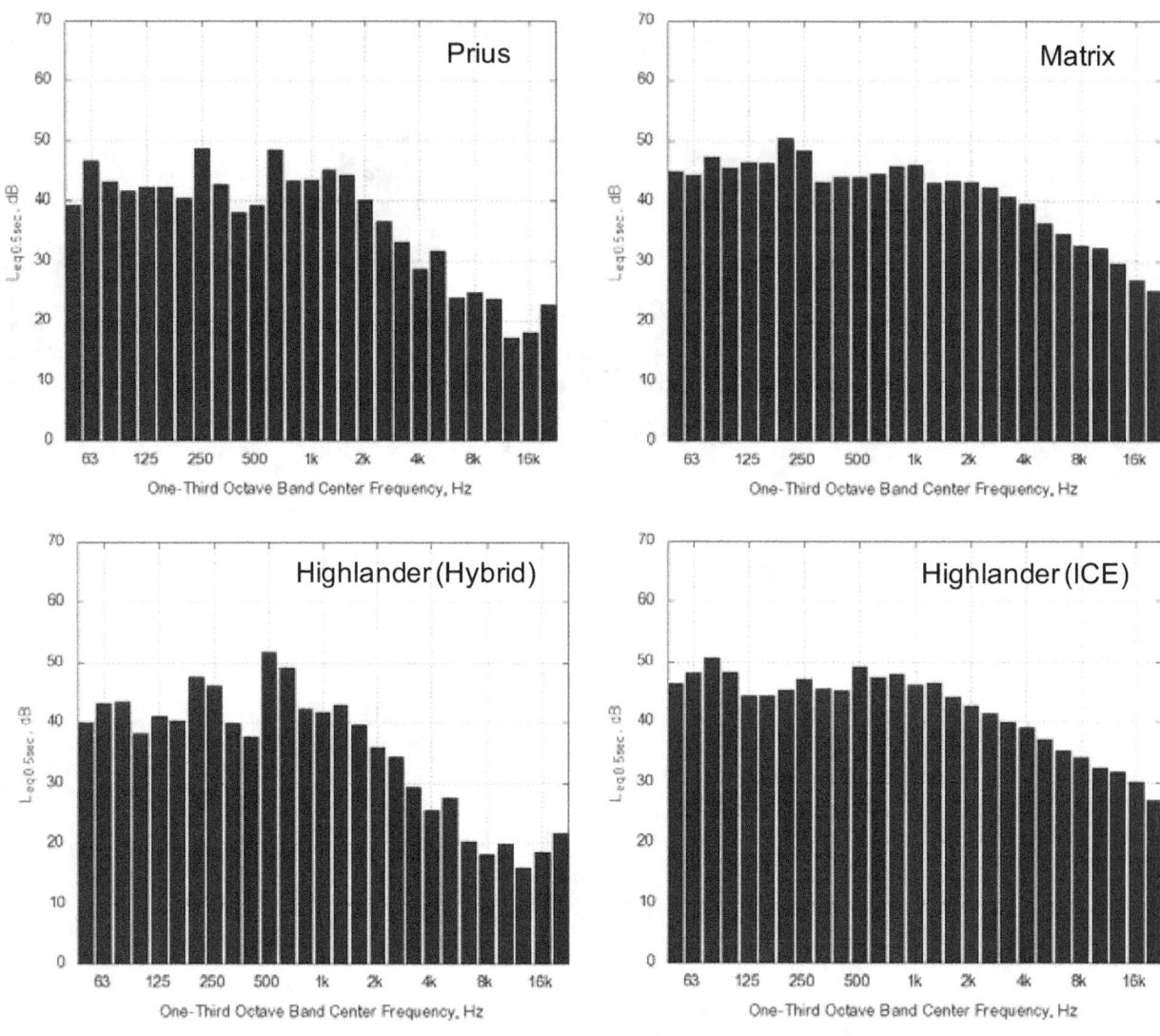

Figure 11. Sample One-Third Octave Band Spectra for Decelerating Passby at 12-ft Microphone Location

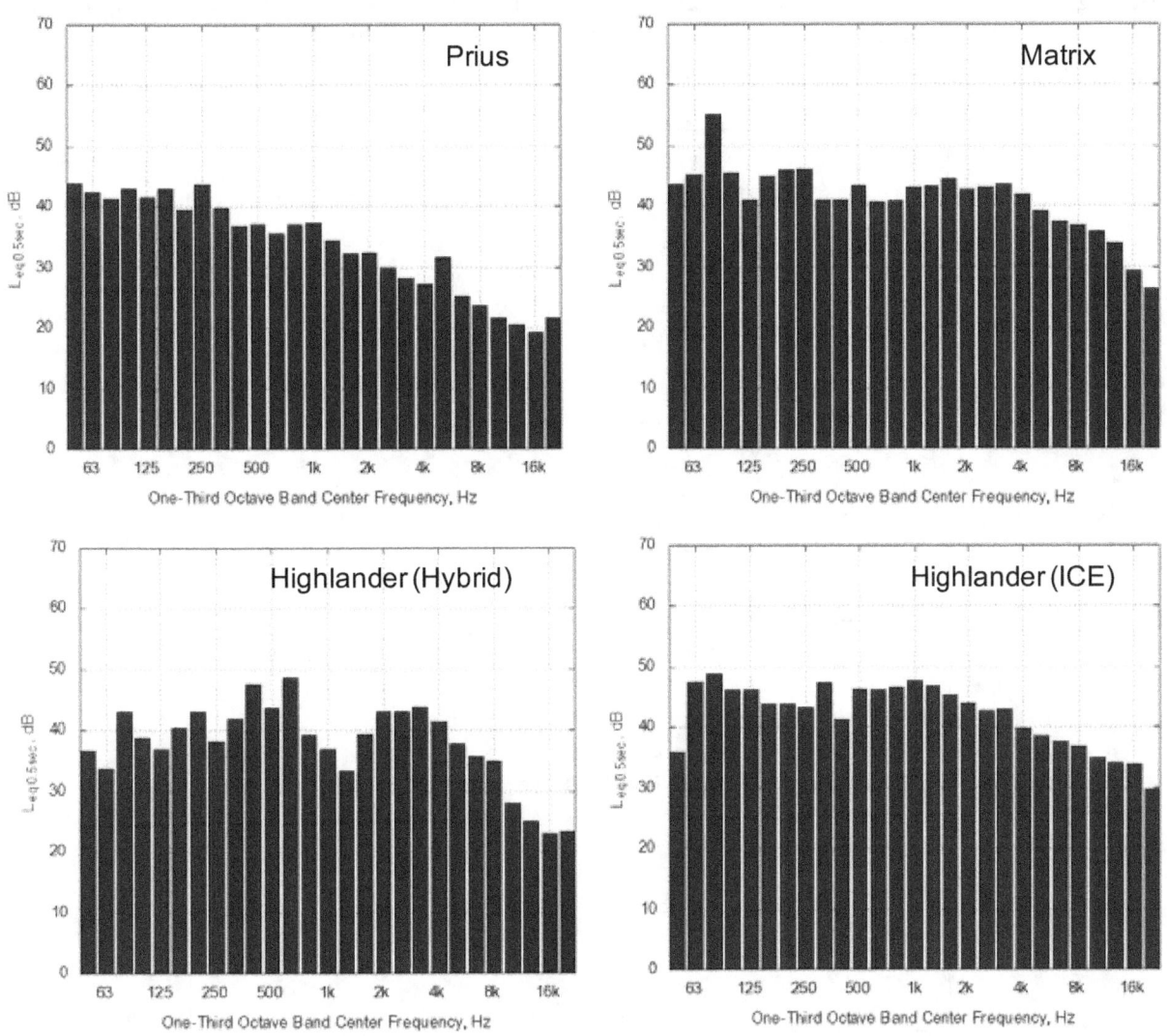

Figure 12. Sample One-Third Octave Band Spectra for 6 mph Constant Speed Passby at 12-ft Microphone Location

4.2.1.4 Level versus Speed at the 12-ft Microphone Location

One question that arises is whether or not the overall sound levels of the two types of vehicles converge at higher speeds due to the dominance of tire noise. In order to document the convergence at higher speeds, overall maximum A-weighted levels at passby are shown in Figure 13 to Figure 14 as a function of speed for the three pairs of vehicles.

The sound level for the Toyota hybrids is 1.4 to 8.8 dB(A) lower than that for the ICE vehicle twins at speeds lower than 10 mph. The Prius converges with the Matrix after 20 mph. The Highlander hybrid converges with the ICE vehicle twin after 10 mph. For both Toyota models, hybrid idle was too low to be accurately measured and is therefore not shown in these figures.

The Honda Civics did not show as great a difference in sound level at low speeds; however, during the experiments it was not possible to get the hybrid Civic to operate in EV-only mode. Therefore, all measurements of the Honda Civic hybrid include engine noise. The Honda Civic hybrid converges with its ICE vehicle twin after 10 mph.

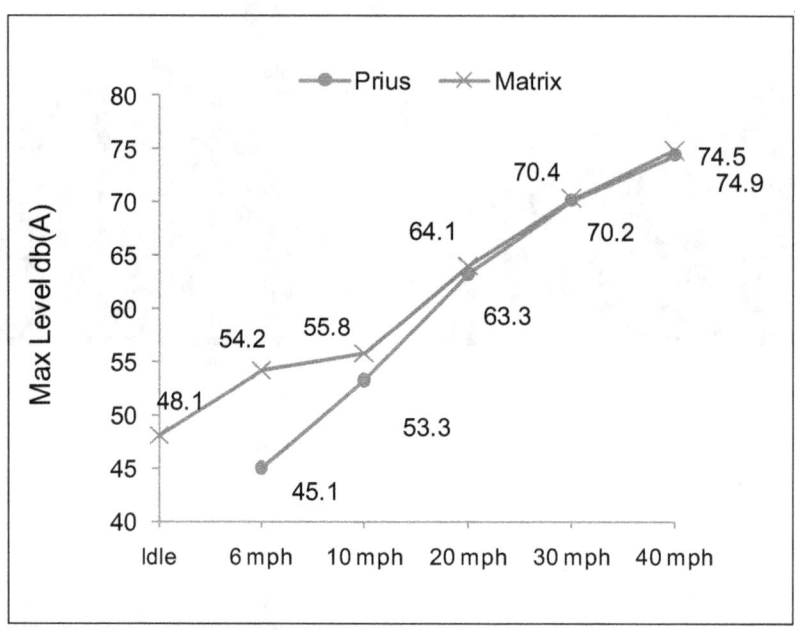

Figure 13. Maximum Levels in dB(A) for the Prius (O) and Matrix (X).

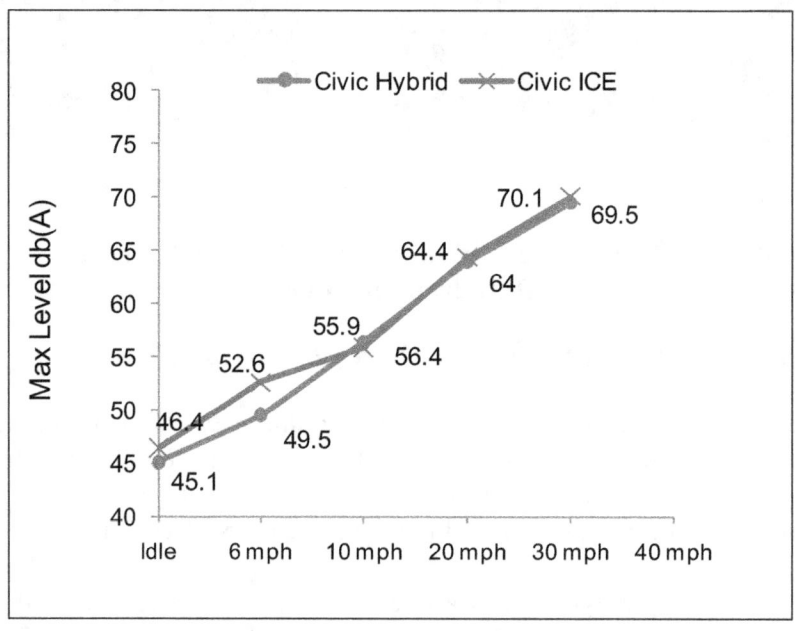

Figure 14. Maximum Levels in dB(A) for the Civic Hybrid (O) and ICE (X).

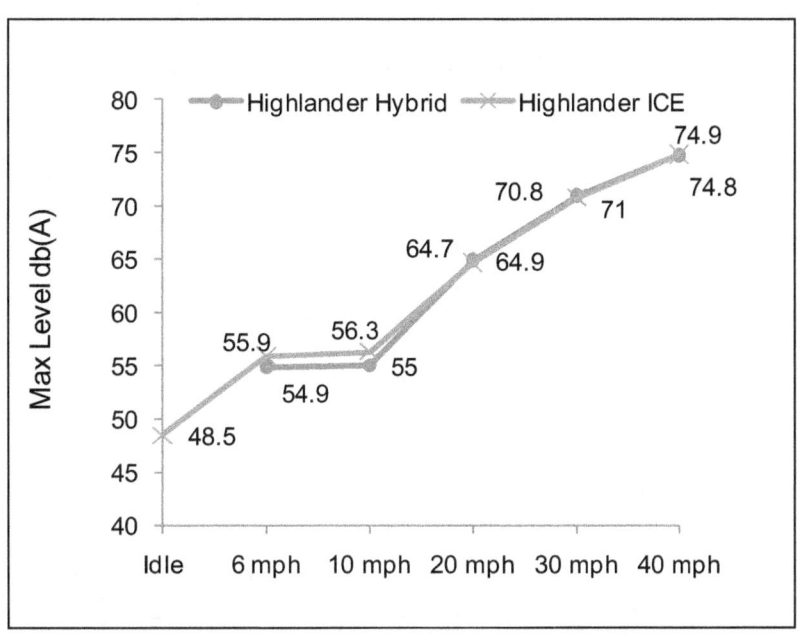

Figure 15. Maximum Levels in dB(A) for the Highlander Hybrid (O) and ICE (X)

4.2.2 Comparison of SAE 6.56-ft with the 12-ft Microphone Test

SAE is developing standard J2889-1 to measure minimum noise emitted by road vehicles. However, because the purpose of J2889-1 is the measurement of minimum noise and the purpose of this study is to evaluate noise emissions on critical safety scenarios, the two approaches are slightly different. The two main differences are that: (1) the SAE standard places a single microphone 6.56 ft (2 m) from the center line at a height of 3.94 ft (1.2 m) above the ground while the most comparable microphone location for this study is 12 ft from the center line at a height of 5 ft above the ground; and (2) the SAE standard considers two operating modes: stationary and 6.2 mph (10 km/h), while this study considers additional speeds, as well as acceleration and deceleration. Measurements were made using the three non-hybrid vehicles (Honda Civic, Toyota Matrix, and Toyota Highlander) using the SAE method while simultaneously measuring 12 ft from the center line and 5 ft above the ground. This activity was completed to document the differences between these two approaches. A comparison of the results is shown in Figure 16 for several stationary configurations; i.e., measuring on the driver side or passenger side, with the vehicle in neutral or park, and with the foot brake applied or not. The difference in sound levels between the two methods was very consistent with the average result of a 3.9 dB higher level for the SAE microphone position. The Toyota Matrix was also tested for 6 mph constant speed passby where the difference between the two methods was 4 dB.

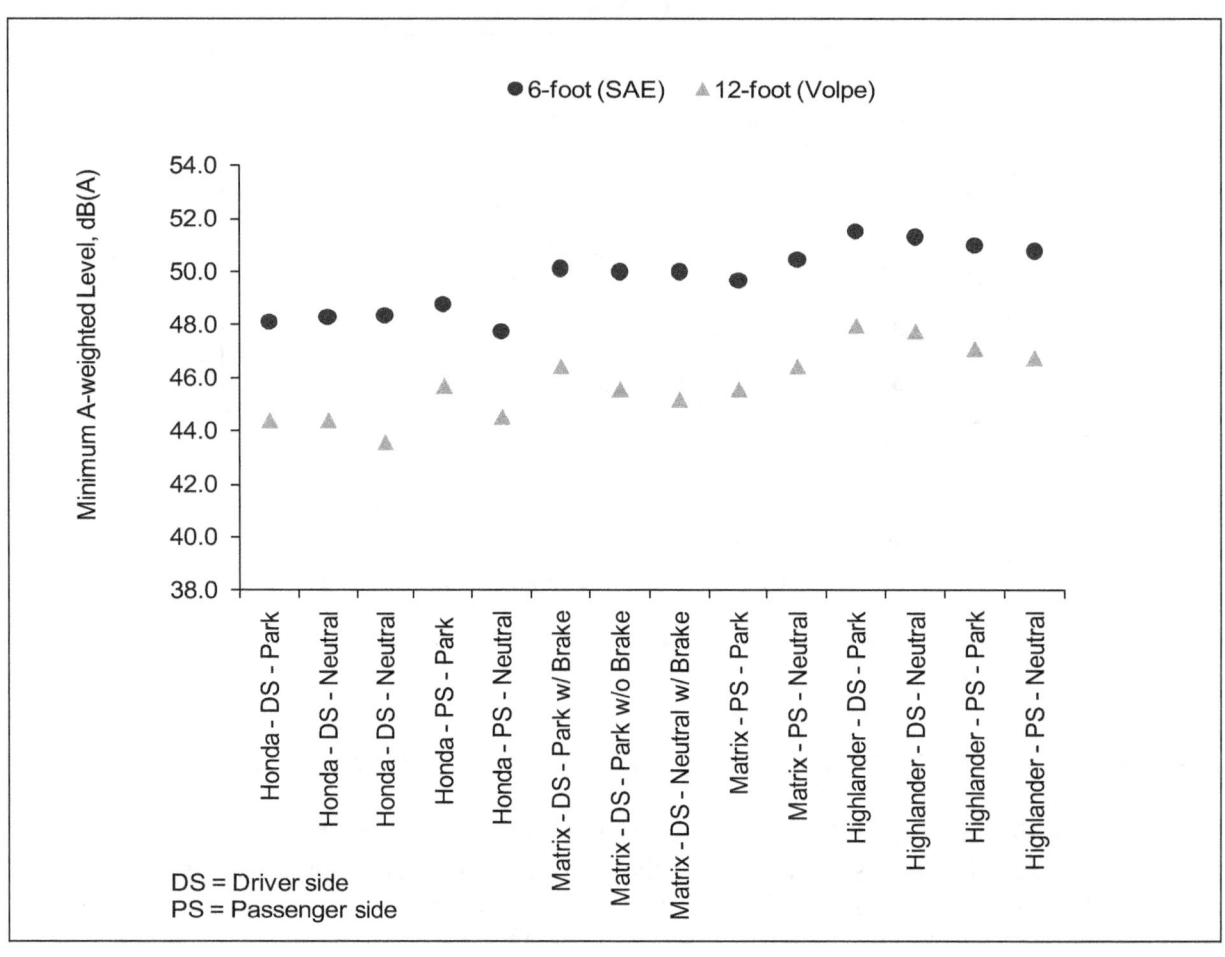

Figure 16. Comparison of Idle Measurements at 6.6 ft (according to SAE J2889-1) and 12 ft (Volpe)

4.3 Ambient Measurement for Critical Safety Scenarios

The purpose of acquiring the binaural ambient audio recordings was to combine this data with vehicle audio recordings for use in the human subject studies. Data was also collected to characterize the ambient noise at the location using an SLM.

Ambient sound measurements were recorded and measured at four sites. Candidate sites were selected in order to provide ambient levels that covered a range of levels from "quiet" to "loud," had sounds that were characteristic of those encountered by pedestrians, and were continuous so that a sudden high ambient level would not mask the vehicle sound just as it was becoming detectable. The four sites evaluated included: Site A (Hunnewell Avenue and Braemore Road, Newton, Massachusetts); Site B (Carroll Center for the Blind); Site C (Perkins School for the Blind); and Site D (TRC). Average A-weighted sound pressure levels were as follows: Site A = 56.2 dB(A); Site B = 49.5 dB(A); Site C = 49.8 dB(A); and Site D= 31.2 dB(A). The average one-third octave band spectrum for each of the sites is shown in Figure 17. Site A was ultimately considered too loud to be representative of an ambient condition where a pedestrian could reasonably consider it safe to cross a street using acoustic cues alone, and was therefore not used for the final human performance testing. Although Sites B and C had similar overall levels, Site C had a more continuous sound pattern, which made it a more suitable site for use in the human subject studies than Site B for a typical ambient condition. Site D had the lowest overall levels, which made it the best candidate to use in the human subject studies for a quiet ambient condition.

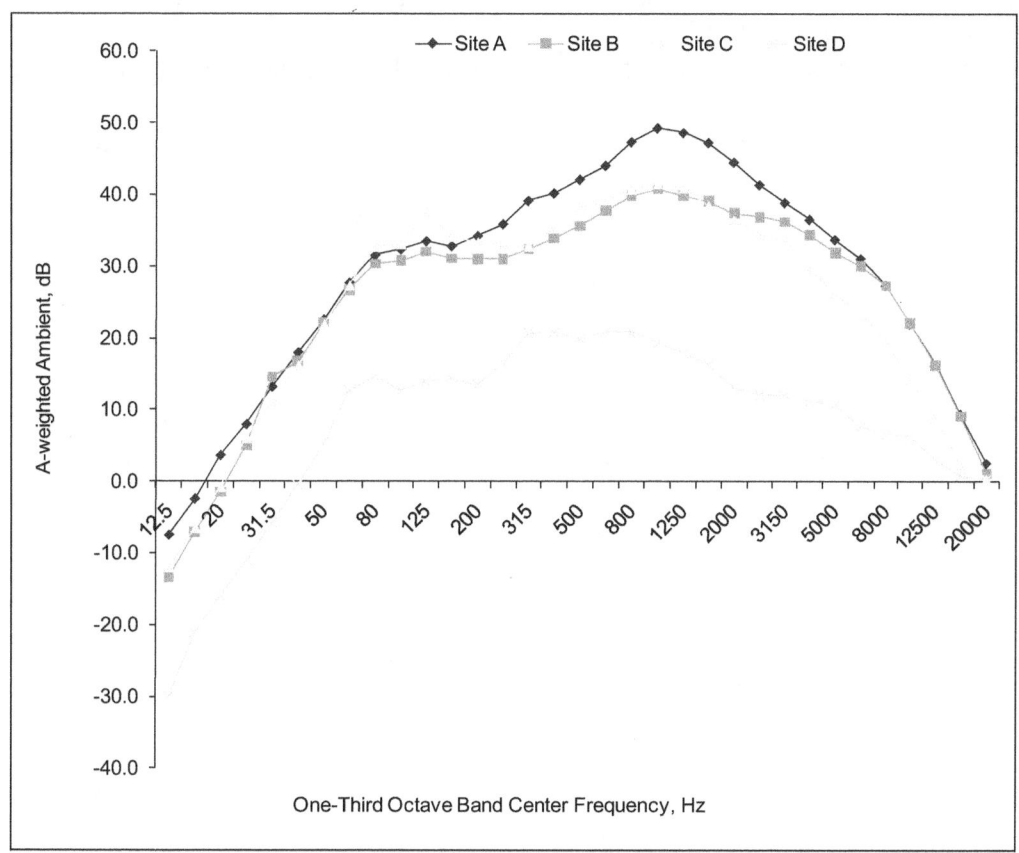

Figure 17. Average A-Weighted One-Third Octave Band Levels for Ambient Measurements: All Sites

4.4 Discussion

This section discusses the acoustic measurement of vehicles under identified critical safety scenarios and the acoustic measurement of ambient levels at several locations considered representative of ambient levels at pedestrian crossings. Data recorded with SLM are reported here and in Appendix A. Binaural recordings for vehicles were combined with ambient recordings for use in human subject studies.

Overall sound levels at low speeds were lower for the HEVs than for the ICE vehicles tested. Overall sound levels for the two types of vehicles converge at higher speeds. The speed at which the vehicles converge varies among the three sets of vehicles tested. ICE vehicles converge with the HEVs tested after 10 mph, except for the Toyota Prius, which converges with the Matrix after 20 mph. Acoustic data was also recorded for vehicles traveling in reverse at 5 mph. In this case the overall sound levels for HEVs are 7 to 10 dB(A) lower than the overall sound levels for ICE vehicles. The overall sound levels for HEVs and ICE vehicles did not differ considerable when slowing from 20 mph to 10 mph. The overall sound levels for HEVs approaching at a constant speed of 6 mph were 2-8 dB(A) lower than for their ICE twins. The sound levels for the Toyota hybrids when stationary were too low to be measured under the ambient condition present. Finally, there is a trend for HEVs to have less high frequency content relative to the overall sound level compared to ICE vehicles. There is an exception to this trend with a notable peak in the Toyota vehicles in the 5 kHz one-third octave band sound level when slowing or braking.

5. AUDITORY DETECTABILITY OF VEHICLES IN CRITICAL SAFETY SCENARIOS

5.1 Procedure to Measure the Auditory Detectability of Vehicles

The study examines whether the auditory detectability differs between ICE vehicles and HEVs operated in electric mode and if so, by how much. Each subject listened to binaural recordings of the sounds of various vehicles in selected scenarios with either a relatively quiet rural ambient sound or with a moderately noisy suburban ambient sound superimposed. The auditory detectability of vehicles in critical safety scenarios is described in this section.

5.1.1 Objectives

This study examines how pedestrian performance (response time and accuracy) for three vehicle maneuvers differs for ICE vehicles and HEVs operated in electric mode. A second objective is to examine how ambient sound levels affect the ability of blind subjects to detect the vehicles. Three types of vehicle maneuvers with high potential for adverse consequences for blind pedestrians were selected for testing: (1) vehicle backing out (e.g., driveway); (2) vehicle approaching at a constant low speed; and (3) vehicles moving in parallel and slowing (as if to turn right).

5.1.2 Subjects

A total of 51 volunteers 18 or older participated in the study; 48 completed all the sessions in the study. Participation was limited to independent travelers who complete street crossing on a regular basis, are legally blind, and self-reported to have normal hearing in both ears without hearing aids. Subjects were recruited among the students, staff, and library users affiliated with the Carroll Center for the Blind in Newton, Massachusetts, and the Perkins School for the Blind in Watertown, Massachusetts. Individuals with severe hearing loss, users of hearing aids, and persons who do not travel independently on a regular basis were not eligible because they introduce significant variance in the detection task and mask the effects of the other independent variables that are the principal focus of this research.

All subjects were briefed by a Volpe Center investigator using the New England Institutional Review Board protocol for the protection of human subjects. Potential subjects were given a description of the experimental set up, the task to be performed, duration of the test, and the test protocol in a recruiting e-mail sent by the Carroll Center for the Blind and the Perkins School for the Blind. Those interested in the test contacted the researchers at the Volpe Center. A researcher administered an eligibility determination questionnaire. An informed consent form was sent to each candidate; electronic and Braille copies were made available. After reviewing the informed consent form, those individuals who judged themselves able to perform the test were included as subjects. Eligible subjects scheduled an interview time of mutual convenience with the Volpe Center staff. All subjects signed the informed consent form after discussing any questions with the investigator. Each subject completed all three sessions (one for each vehicle maneuver); a total of 144 observations were recorded for each subject. Participants received compensation for their time in the form of a gift card ($50).

Forty-eight subjects completed the study (46% male and 54% female). Subject ages ranged from 18 to 69 years old. Table 10 shows the age distribution for subjects who participated in the study. The distribution of subjects by type of vision loss is shown in Table 11. Participants in the study are all legally blind. The group includes individuals who are totally blind, blind with light perception, and blind with some usable vision. Subject distribution by mobility aid use is shown in Table 12. Sixty-three percent of the subjects currently use white canes as the primary mobility aids, 33 percent use guide dogs, and 4 percent do not use canes or guide dogs.

Table 10. Subject Age Distribution

Age Group	Number of Subjects	% of All Subjects
18-19	5	10
20-29	8	17
30-39	2	4
40-49	10	21
50-59	18	38
60-69	5	10
TOTAL	48	100

Table 11. Subject Vision Loss Category

Vision Loss Category	Number of Subjects	% of All Subjects
Totally Blind	20	42
Blind–Light Perception	10	21
Blind–Some Usable Vision	18	38
TOTAL	48	100

Table 12. Subject Mobility Aid Usage

Mobility Aid	Number of Subjects	% of All Subjects
White Cane	30	63
Guide Dog	16	33
None	2	4
TOTAL	48	100

5.1.3 Apparatus

The experiment was constructed using the E-Prime system from Psychology Software Tools, Inc. The study setup includes include a Toshiba mini laptop computer used by the investigator to run the test; two sets of headphones, one for the subject (Grado Labs Prestige Series SR125) and one for the investigator (Sony Dynamics Stereo MDR-V6); and a full size computer keyboard. Headphones were sanitized after each participant usage. In each trial, subjects heard a binaurally recorded audio file exactly 12 seconds long and terminating at the moment the vehicle passed the microphone. The sound produced at the subjects' ears matched the sounds measured at the time the vehicles and ambient sounds were recorded. The playback system was calibrated and the computer volume fixed prior to data collection. The experimental sessions were conducted on the campuses of the Perkins School for the Blind and the Carroll Center for the Blind. Each institution provided a quiet space for the experiment and assistance in recruiting participants.

5.1.4 Study Design and Methods

The study consisted of three sessions—one for each of the following vehicle maneuvers:

1. Vehicle backing out at 5 mph from the left (e.g., driveway);
2. Vehicle approaching a constant low speed of 6 mph from the left; and
3. Vehicle moving in parallel and slowing from 20 to 10 mph (as if to turn right).

The sessions were structured to examine the effect of ambient sound level and vehicle type for each vehicle maneuver. In particular, detectability was examined in two ambient sound levels: one relatively quiet rural ambient with overall sound level of 31.2 dB(A) and a moderate noisy suburban ambient sound with overall sound level of 49.8 dB(A). Four vehicles were included in the test: two hybrid electric operated in electric mode (Toyota Prius and Highlander) and two ICE vehicles (Toyota Matrix and Highlander). These four vehicles were selected from those vehicles recorded based on the potential of HEVs to operate in electric mode. The Honda Civic recorded does not operate in electric-only mode. Table 13 shows the test conditions and the levels of each of the variables.

Table 13. Test Conditions for Human Subject Studies

Conditions	Levels
Vehicle Maneuver (Traveling Situation)	Vehicle backing out of a driveway at 5 mph from the left Vehicle approaching a constant low speed of 6 mph from the left Vehicle moving in the parallel street, but slowing down from 20 to 10 mph as if to turn right
Ambient Sound Level	TRC rural ambient sound (low) = 31.2 dB(A) Suburban ambient sound (high) = 49.8 dB(A)
Vehicles	Prius in electric mode (HE) Matrix (ICE vehicle) Highlander in electric mode (HE) Highlander (ICE vehicle)

Subjects were briefed individually according to the informed consent form required by the IRB. The Volpe Center investigator answered any questions from the subject and both parties signed the informed consent form. The investigator described the traveling situation before each session. Each session began with eight practice trials containing examples of the sounds of all of the target vehicles as well as examples with no target vehicle present. The practice session allowed subjects to experience the relatively uniform ambient, environmental sounds, as well as to familiarize themselves with the traveling situation and experimental task. The investigator provided feedback during the practice session about whether the subject was making the correct response. Feedback was not given during the experimental trials.

Each experimental session included 48 recordings or trials with each trial consisting of the combination of the sound of a particular vehicle (either ICE vehicle, HEV, or no target vehicle) executing the maneuver in question and a particular ambient sound (either quiet rural or moderate suburban). Each of the sessions had two blocks, one for each ambient sound level. Each block consisted of 24 trials. A grand total of 144 observations per subject were recorded.

The investigator listened to the recording throughout the experiment to ensure the program was running correctly. Each combination of vehicle maneuver and ambient was repeated six times (four times with and two times without target vehicles). The no-signal condition (e.g., target vehicle not present) was needed to determine the percentage of correct detection and more importantly the frequency of misses or incorrect detection. Subjects were instructed to press the computer space bar only when and if they first heard a target vehicle. If they did not hear the target vehicle (any vehicle in the backing and side-approach sessions or a decelerating vehicle in the right-turn session), they were instructed not to press the space bar and to wait for the sound clip to end and for the next one to begin. There was a 5-second average transition between trials.

Due to the variability between subjects, all subjects were presented with the same experimental trials (within-subject design). A total of 144 observations were recorded for each subject. The presentation order for the vehicle maneuvers and ambient level was counterbalanced across subjects. The presentation of vehicle/no vehicle trials and vehicle types were randomly distributed within subjects.

A total of 1 hour and 15 minutes per subject was required to complete the study. Time was reserved to discuss the test protocol, to complete practice trials, to answer any questions the subject may have before the experiment, and to provide breaks between sessions. Subjects were tested one at a time with the investigator seated next to the subject. For each session the Volpe Center investigator explained the overall procedure as well as the traveling situation (vehicle maneuver) being tested and how to use the response key (i.e., computer space bar). There was a pre-study questionnaire and a debriefing at the end.

5.1.5 Data Reduction

Performance measures include: missed detection (a target vehicle is present and the subject missed it); response time (time elapsed from the start of the trial to the moment the subject pressed the space bar as an indication he/she detected the target vehicle); time-to-vehicle-arrival (time from first detection of a target vehicle to the instant the vehicle passes the microphone line/pedestrian location); and detection distance (distance between the vehicle and the microphone/pedestrian location at the moment the subject indicates detection). Detection distance for a vehicle approaching at a constant-speed is calculated by multiplying the time-to-vehicle-arrival by the vehicle speed (expressed in feet per second). Detection distance for the "slowing vehicle" scenario is calculated according to the following equation:

$$d = (v_f t) + (½ a t^2)$$

where: d = distance at which detection occurred
v_f = velocity at microphone line (i.e., 10 mph or 14.67 ft per sec)
t = time-to-vehicle-arrival (i.e., seconds until vehicle passed microphone)
a = deceleration rate (i.e., 1 m/sec² or 3.28 ft/sec²)

The time-to-vehicle-arrival and detection distances are the basis for evaluating the safety risk by vehicle and ambient sound for each vehicle maneuver.

5.2 Results of Human Subject Studies

Results of the human subject tests are presented for each vehicle maneuver or traveling situation: vehicle backing up; vehicle approaching in parallel and slowing; and vehicle approaching at a constant low speed. Three independent variables are considered: vehicle maneuver, vehicle type, and ambient sound level. Two dependent variables are examined: missed detection frequency and response time. Missed detection frequency is defined as the instances when the target vehicle was present and the subject failed to respond.

Response time is computed as the time from the start of the trial to the instant the subject pressed the space bar as an indication he/she detected the target vehicle. The difference between the trial duration and the response time gives the time-to-vehicle-arrival. For example, a response time of 7 seconds is associated with a time-to-vehicle-arrival of 5 seconds (12 minus 7), which means the pedestrian detected the target vehicle 5 seconds before the vehicle arrived at the microphone/pedestrian location. In terms of safety and collision avoidance high response times and small time-to-vehicle-arrival could be dangerous. A repeated measures analysis of variance (ANOVA) is used to analyze the main and interaction effects of the independent variables; vehicle type, vehicle maneuver and ambient sound level. Considering all three independent variables, there is a main effect of vehicle type [$F (2.5, 119.4) = 78.13$; $p < 0.05$], vehicle maneuver [$F (1.69, 79.59) = 146.49$; $p < 0.05$], and ambient sound level [$F (1, 47) = 94.21$; $p < 0.05$]. Similarly, there are interaction effects between vehicle type and ambient [$F (2.68, 125.89) = 4.54$; $p < 0.05$]; vehicle type and maneuver [$F (3.818, 179.43) = 137.37$; $p < 0.05$], ambient and vehicle maneuver [$F (1.99, 93.31) = 31.71$; $p < 0.05$], and a three way interaction between ambient, vehicle type and vehicle maneuver [$F (4.6, 216.50) = 9.673$; $p < 0.05$]. A pair-wise t-test compares each vehicle with the other (ICE vehicle and HEV twins) for each ambient sound level. The results for each of the three vehicle maneuvers are reported below.

5.2.1 Vehicle Backing out

5.2.1.1 Missed Detection

This analysis includes data for all 48 subjects in the "vehicle backing out" session. Each subject received 48 trials. A target vehicle was present in 32 trials; 16 for each ambient sound condition. Subjects were more likely to miss the Toyota HEVs than the Toyota ICE vehicles in the backing out session.

Figure 18 shows the missed detection rate for the "vehicle backing out" session. The missed detection rate is computed as the total number of trials where subjects missed a target vehicle, divided by the total number of trials with a target vehicle present (for all subjects). The missed detection rates in the low ambient condition are: 0.05 for the Prius; 0.02 for the Matrix; 0.10 for the Highlander Hybrid; and 0.02 for the Highlander ICE. The corresponding values for the backing out scenario in the high ambient condition are: 0.11 for the Prius; 0.0 for the Matrix; 0.26 for the Highlander; and 0.02 for the Highlander ICE.

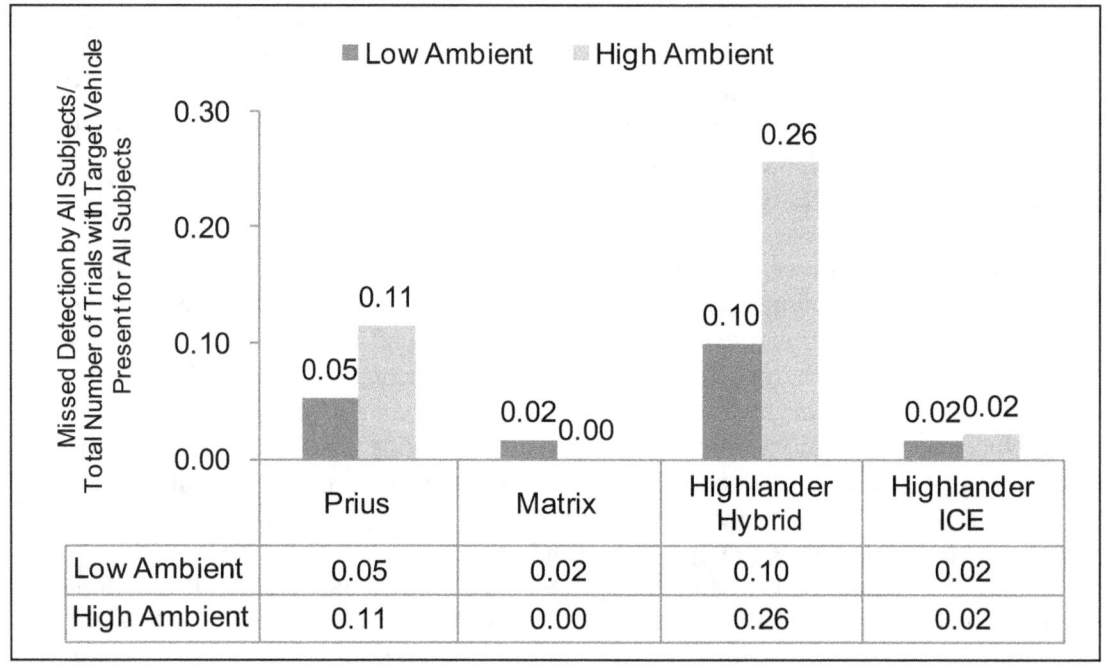

Figure 18. Missed Detection Rates for the "Vehicle Backing Out" Scenario

Twenty-three subjects failed to detect a vehicle at least once during the session. Nineteen of the 23 subjects failed to detect an HEV at least once, and four subjects failed to detect both an HEV and an ICE vehicle at least once. Figure 19 shows the distribution of subjects by the number of missed detections for each vehicle and ambient condition. Five subjects failed to detect one or more vehicles in given ambient sound conditions. All five of these failed to detect the Toyota Highlander Hybrid and three of these also failed to detect the Toyota Prius.

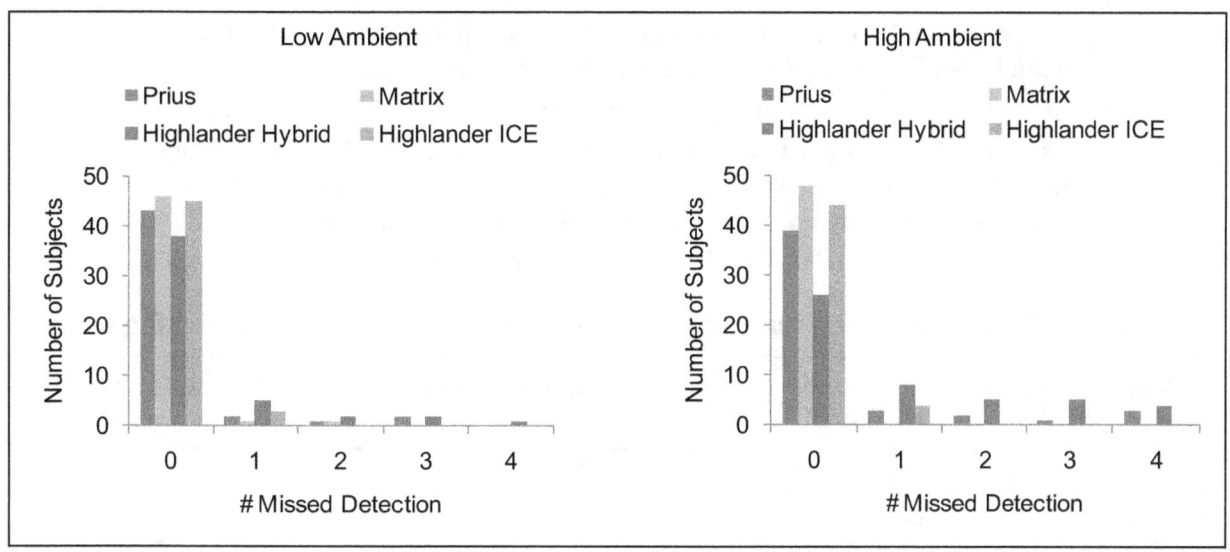

Figure 19. Distribution of Subjects by the Number of Missed Detection: Vehicles Backing Out

5.2.1.2 Response Time and Time-to-Vehicle-arrival

Mean response times and corresponding time-to-vehicle-arrival are shown in Table 14 for the "vehicle backing out scenario." Results are for the 48 subjects that completed the study. Each subject received 32 trials with a target vehicle present, 16 for each ambient condition, 4 for each vehicle. The maximum possible value for response time was assigned when a subject failed to detect a target vehicle. The maximum value for response time is equal to the duration of the trial (i.e., 12 seconds).

The main effect of ambient and vehicle are statistically significant. Figure 20 shows a comparison of the time-to-vehicle-arrival for each vehicle-ambient combination. (Note: two bars within an ambient condition are significantly different if they have a different letter at their base.)

- Subjects took longer to detect vehicles in the high ambient sound condition than in the low ambient sound condition. The main effect of ambient is statistically significant [$F(1, 47) = 96.64$; $p < 0.05$, Greenhouse-Geisser correction for sphericity].
- The average response time is 7.6 seconds for the low ambient and 9.3 seconds for the high ambient. These correspond to a time-to-vehicle-arrival of 4.4 and 2.7 seconds for the low and high ambient condition, respectively.
- Subject took longer to detect both HEV than their ICE twins. The main effect of vehicle is statistically significant [$F(2.72, 128.0) = 115.0$; $p < 0.05$, Greenhouse-Geisser correction for sphericity].
- In the low ambient condition, subjects detected both ICE vehicles sooner than their HEV twins. The Toyota Matrix was detected 1.2 seconds sooner than the Toyota Prius. The Toyota Highlander ICE was detected 1.9 seconds

sooner than the Toyota Highlander hybrid. The differences between ICE vehicles and their HEV twins are statistically significant.

- In the high ambient condition, subjects detected both ICE vehicles sooner than their HEV twins. The Toyota Matrix was detected 1.1 seconds sooner than the Toyota Prius. The Toyota Highlander ICE was detected 1.9 seconds sooner than the Toyota Highlander hybrid. The differences between ICE vehicles and their HEV twins are statistically significant.

Table 14. Mean Response Time and Time-to-Vehicle-arrival: Vehicles Backing Out

Ambient Sound - Vehicle	Response Time (milliseconds)	Standard Deviation (milliseconds)	Time-to-Vehicle-Arrival (seconds)
Low-Prius	8022	1798	4.0
Low-Matrix	6846	1626	5.2
Low-Highlander Hybrid	8657	1735	3.3
Low-Highlander ICE	6827	1714	5.2
High-Prius	9489	1388	2.5
High-Matrix	8379	1065	3.6
High-Highlander Hybrid	10562	1254	1.4
High-Highlander ICE	8691	1212	3.3

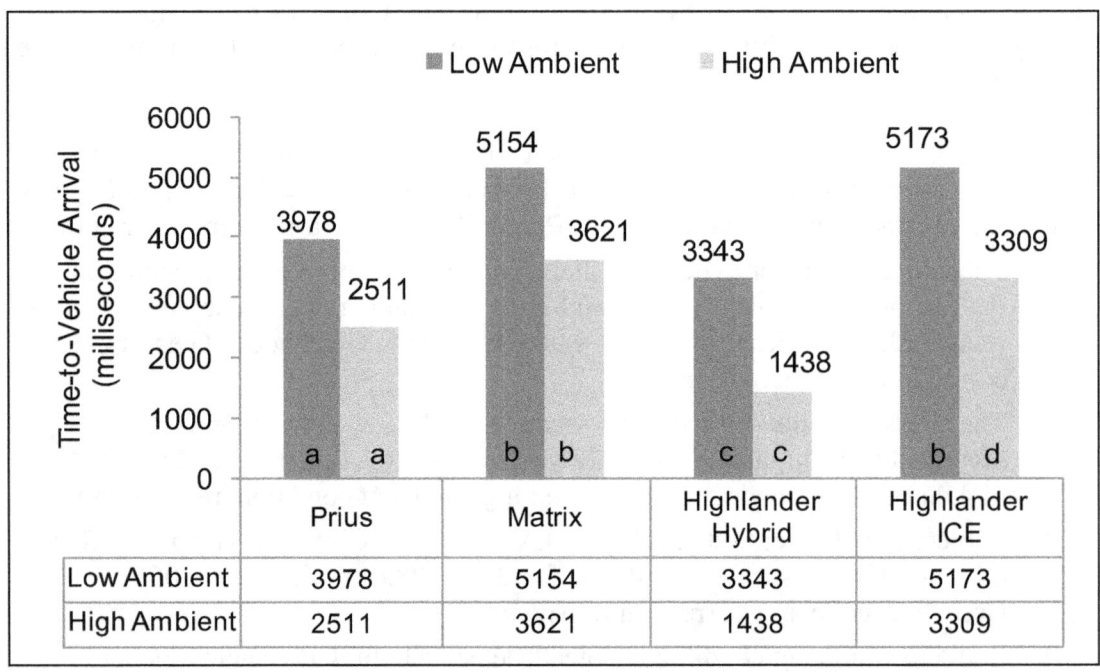

Figure 20. Mean Time-to-Vehicle-Arrival by Vehicle and Ambient Condition: Vehicles Backing Out

5.2.2 Vehicle Approaching in Parallel and Slowing

In the results reported below, it is clear that HEVs are being detected sooner and with better accuracy than ICE vehicles. To understand this anomalous result, one must be aware the Toyota HEVs tested emit a faint 5-kHz tone with a 10-kHz harmonic when they are operating in regenerative braking mode. Figure 21 shows a narrow-band spectral analysis of the sound emitted by a Prius during the slowing maneuver.

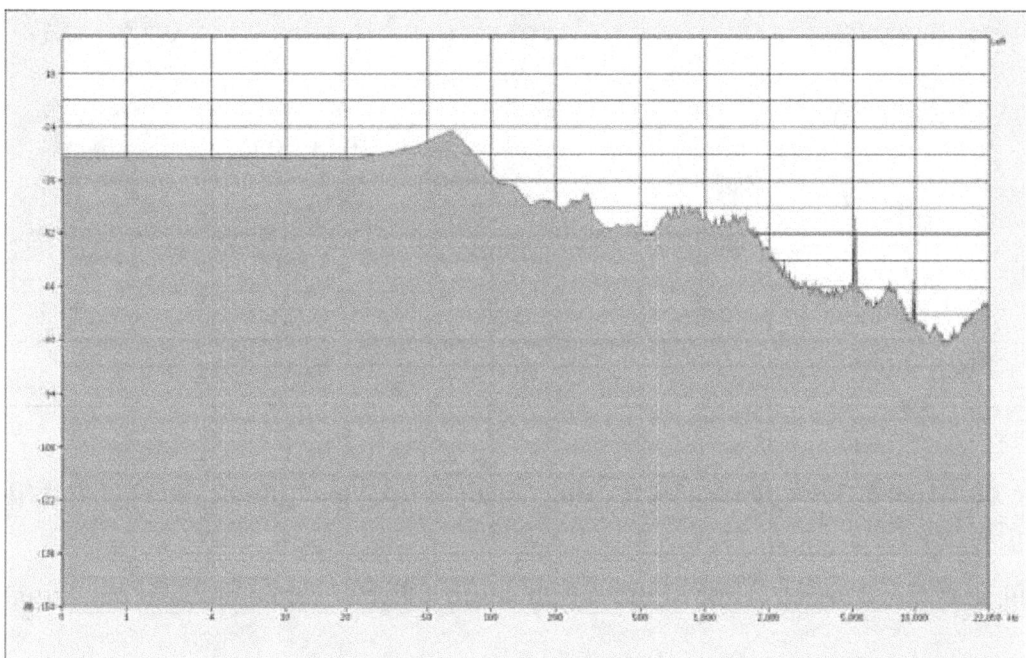

Figure 21. Narrow-Band Spectral Analysis of Prius Braking Sound

Even though the power of the 5-kHz braking tone is miniscule compared with the noise at all other frequencies, it is nonetheless audible. Its presence made it relatively easy for subjects to perceive a braking vehicle as opposed to non-braking vehicle. It was more difficult to make this distinction with ICE vehicles, because the characteristic sound of a slowing ICE vehicle did not become perceptible until roughly one second later in the approach than when the 5-kHz regenerative-braking tone became audible.

5.2.2.1 Missed Detection

The analysis includes data for all 48 subjects in the "vehicle approaching in parallel and slowing" session. Each subject received 48 trials. A target vehicle was present in 32 trials, 16 for each ambient sound condition. Subjects were more likely to miss the Toyota ICE vehicles approaching in parallel lane and slowing than the Toyota HEVs in the same situation. Figure 22 shows the missed detection rates for the "vehicle approaching in parallel and slowing" session. The missed detection rates in the low ambient condition are: 0.05 for the Prius; 0.31 for the Matrix; 0.03 for the Highlander Hybrid; and 0.17 for the Highlander ICE vehicle. The missed detection rates in the high ambient condition are: 0.05 for the Prius; 0.35 for the Matrix; 0.03 for the Highlander Hybrid; and 0.17 for the Highlander ICE vehicle.

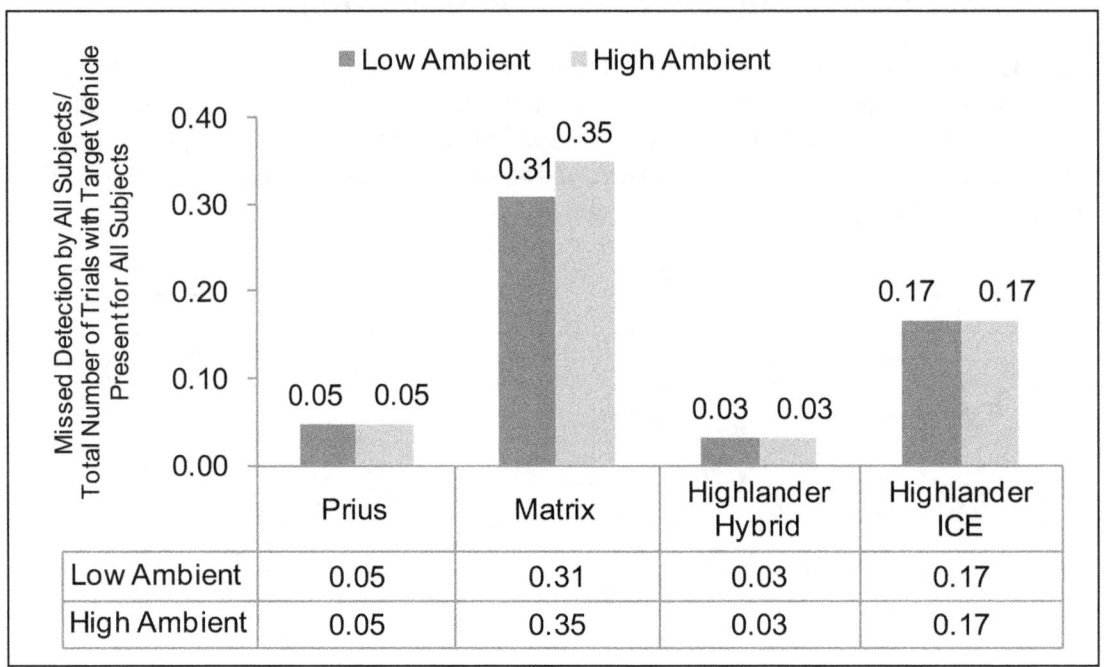

Figure 22. Missed Detection Rates for the "Vehicle Approaching in Parallel and Slowing" Scenario

Eight subjects failed to detect one or more vehicles in a given vehicle-ambient condition. Of these:
- 5 subjects failed to detect the Matrix;
- 1 subject failed to detect the Highlander ICE;
- 1 subject failed to detect, the Prius, Matrix and the Highlander ICE; and
- 1 subject failed to detect the Highlander hybrid and the Highlander ICE.

Figure 23 shows the distribution of subjects by the number of missed detection for each vehicle and ambient condition.

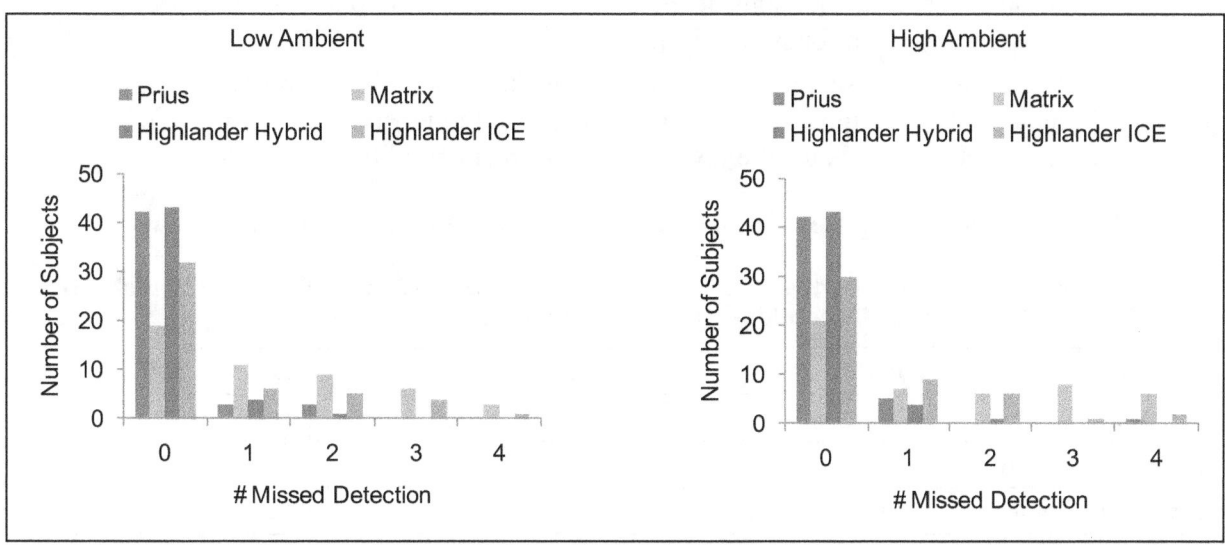

Figure 23. Distribution of Subjects by the Number of Missed Detection: Vehicle Slowing

5.2.2.2 Response Time and Time-to-Vehicle-Arrival

Mean response times and mean times-to-vehicle-arrival for the vehicle slowing scenario are shown in Table 15. The results are for the 48 subjects that completed the study. Each subject received 32 trials with a target vehicle present, 16 for each ambient condition, 4 for each vehicle. The maximum possible value for response time was assigned when a subject failed to detect a target vehicle. The maximum value for response time is equal to the duration of the trial (i.e., 12 seconds). The main effect of ambient and vehicle are statistically significant. Figure 24 shows ambient comparison of the time-to-vehicle-arrival data for each vehicle-ambient combination. (Note: two bars within an ambient condition are significantly different if they have a different letter at their base.)

- Subjects detected HEVs sooner than their ICE vehicle twins. The main effect of vehicle is statistically significant [$F (2.04, 96) = 163.85$; $p < 0.05$, Greenhouse-Geisser correction for sphericity].

- In the low ambient condition, subjects detected both HEVs sooner than their ICE vehicle twins. The Toyota Prius was detected 0.9 seconds sooner on average than the Toyota Matrix. The Toyota Highlander Hybrid was detected 1.5 seconds sooner on average than the Toyota Hybrid ICE. The differences between HEVs and their ICE vehicle twins are statistically significant.

- In the high ambient condition, subjects detected both HEVs sooner than their ICE vehicle twins. The Toyota Prius was detected 1.1 seconds sooner on average than the Toyota Matrix. The Toyota Highlander Hybrid was detected 1.4 seconds sooner on average than the Toyota Hybrid ICE. The differences between HEVs and their ICE vehicle twins are statistically significant.

- Subjects took slightly longer to detect vehicles in the high ambient sound condition than in the low ambient sound condition. The main effect of ambient is statistically significant [$F(1, 47) = 10.56$; $p < 0.05$, Greenhouse-Geisser correction for sphericity].

- The mean response time is 10.1 seconds for the low ambient and 10.3 seconds for the high ambient. These correspond to mean times-to-arrival of 1.9 and 1.7 seconds for the low and high ambient condition, respectively.

Table 15. Mean Response Time and Time-to-Vehicle-arrival: Vehicle Slowing

Ambient Sound-Vehicle	Response Time (milliseconds)	Standard Deviation (milliseconds)	Time-to-Vehicle-Arrival(s)
Low-Prius	10018	625	2.0
Low-Matrix	10948	628	1.1
Low-Highlander Hybrid	9020	674	3.0
Low-Highlander ICE	10530	812	1.5
High-Prius	10065	563	1.9
High-Matrix	11199	595	0.8
High-Highlander Hybrid	9287	752	2.7
High-Highlander ICE	10670	613	1.3

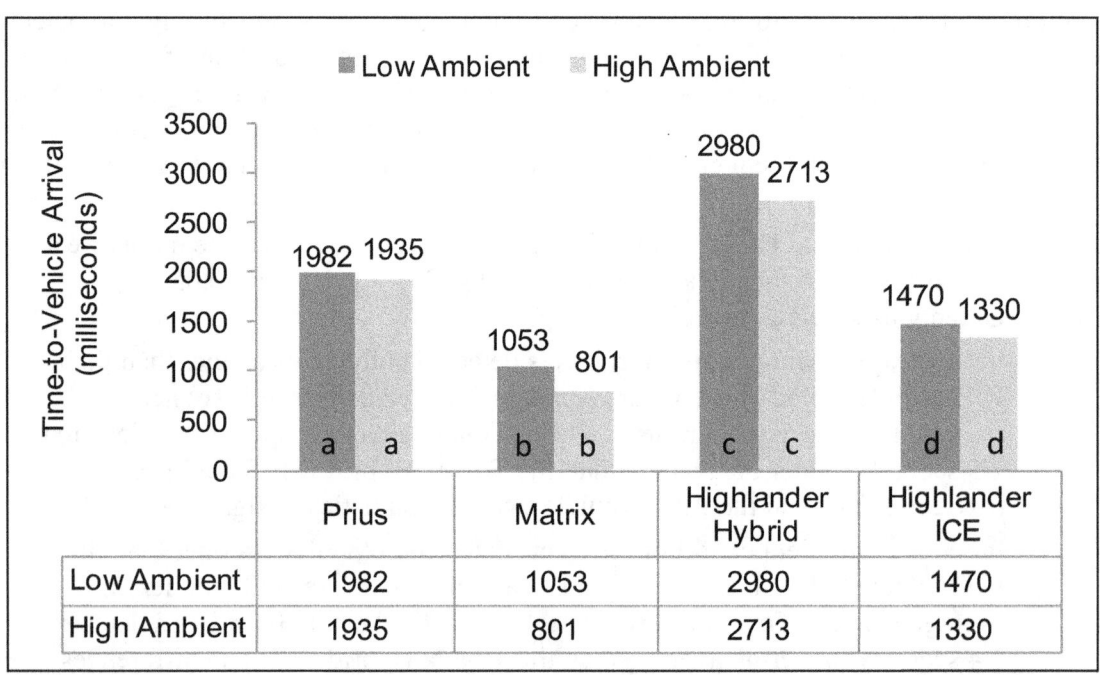

Figure 24. Mean Time-to-Vehicle-Arrival: Vehicle Slowing

5.2.2.3 Detection Distance

Detection distances for the "slowing vehicle" scenario in the low ambient condition are computed as 35.9 ft for the Prius and 18.1ft for the Matrix; 58.8 ft for the Highlander hybrid and 25.7 ft for the Highlander ICE vehicle. The corresponding values in the high ambient conditions are: 33.8 ft for the Prius and 12.8 ft for the Matrix; 51.6 ft for the Highlander hybrid and 21.8 ft for the Highlander ICE vehicle.

5.2.3 Vehicle Approaching at a Constant Low Speed

5.2.3.1 Missed Detection

The analysis includes data for all 48 subjects in the "vehicle approaching at a constant low speed" session. Each subject received 48 trials. A target vehicle was present in 32 trials; 16 for each ambient sound condition. Subjects were more likely to miss the Toyota HEVs than the Toyota ICE vehicles approaching at a constant low speed. Figure 25 shows the missed detection rate for the "vehicle approaching at a constant low speed" session. The missed detection rate is computed as the total number of trials where subjects missed a target vehicle, divided by the total number of trials with a target vehicle present (for all subjects). The missed detection rates in the low ambient condition are: 0.02 for the Prius; 0.01 for the Matrix; 0.03 for the Highlander Hybrid; and 0.0 for the Highlander ICE vehicle. The corresponding values in the high ambient condition are: 0.21 for the Prius; 0.02 for the Matrix; 0.04 for the Highlander; and 0.01 for the Highlander ICE vehicle.

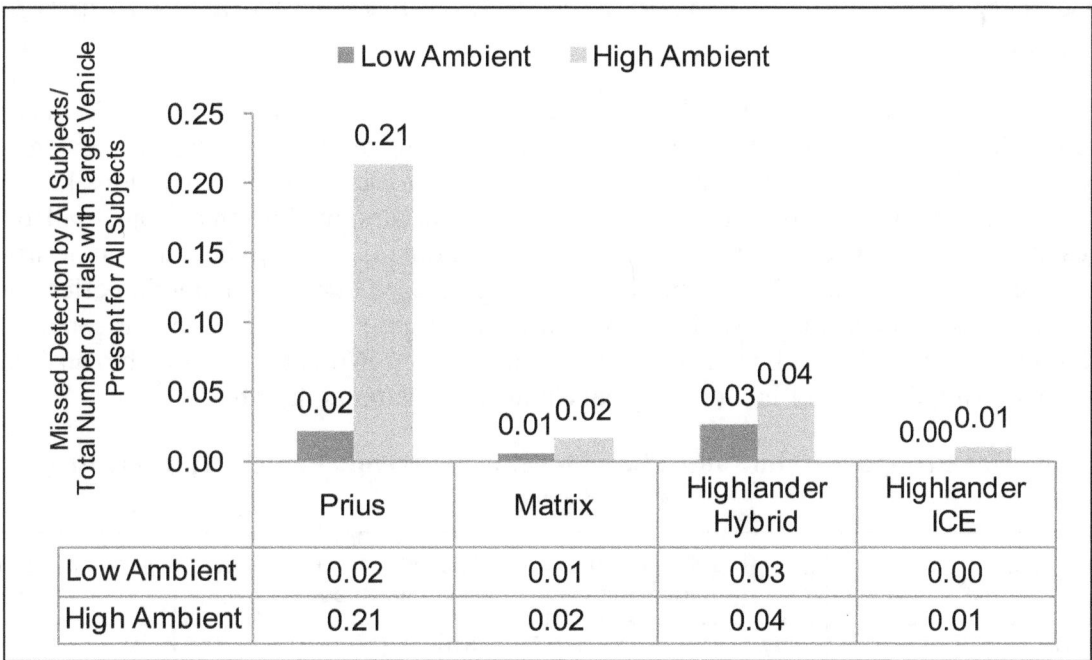

Figure 25. Missed Detection Rates for the "Vehicle Approaching at a Constant Speed" Scenario

Twenty-six subjects failed to detect a vehicle at least once during the session. Twenty-three of the 26 subjects failed to detect a hybrid vehicle at least once and 3 subjects failed to detect both a hybrid and an ICE at least once. Two subjects never detected a Toyota Prius in the high ambient condition. Figure 26 shows the distribution of subjects by the number of missed detection for each vehicle and ambient condition.

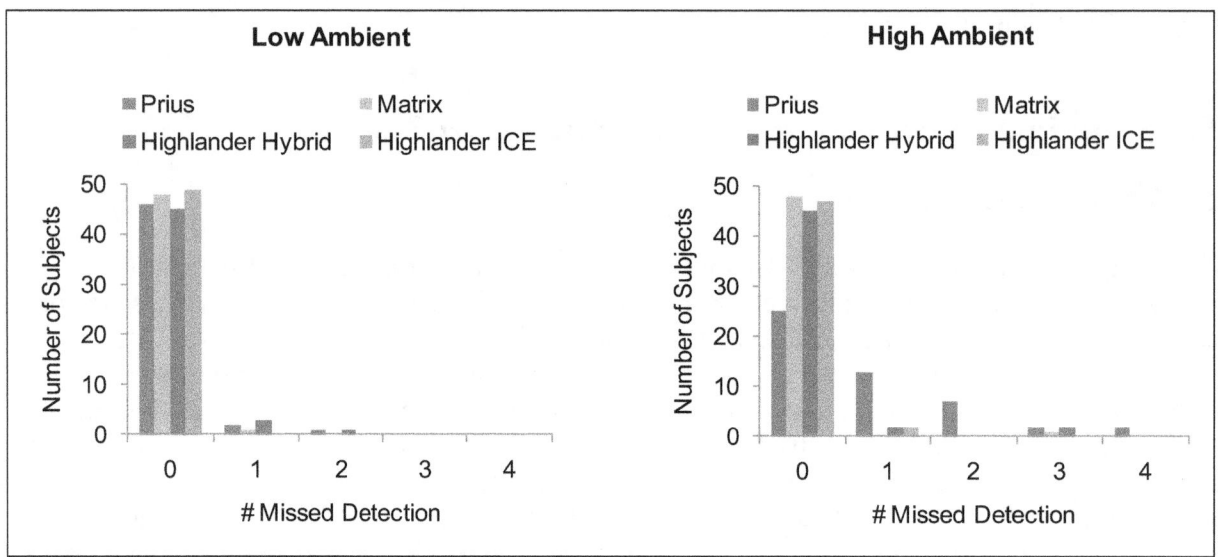

Figure 26. Distribution of Subjects by the Number of Missed Detection: Low Speed

5.2.3.2 Response Time and Time-to-Vehicle-Arrival: Vehicle Approaching at Low Speed

The mean response times for each ambient-vehicle combination are shown in Table 16. This analysis includes data for all 48 subjects in the "vehicle approaching at low speed" session. Each subject received 32 trials with a target vehicle present, 16 for each ambient condition, 4 for each vehicle. The maximum possible value for response time was assigned when a subject failed to detect a target vehicle. The maximum value for response time is equal to the duration of the trial (i.e., 12 seconds). The main effect of ambient and vehicle, as well as the interaction effect between ambient and vehicle are statistically significant. Figure 27 shows a comparison of the time-to-vehicle-arrival for each vehicle-ambient combination. (Note: two bars within an ambient condition are significantly different if they have a different letter at their base.)

Table 16. Mean Response Time and Time-to-Vehicle-Arrival: Vehicle Approaching at Low Speed

Ambient Sound-Vehicle	Response Time (milliseconds)	Standard Deviation (milliseconds)	Time-to-Vehicle-Arrival(s)
Low-Prius	7697	2010	4.3
Low-Matrix	6504	2292	5.5
Low-Highlander Hybrid	6699	2068	5.3

Low-Highlander ICE	5246	2359	6.8
High-Prius	9622	1912	2.4
High-Matrix	7400	1633	4.6
High-Highlander Hybrid	7876	1416	4.1
High-Highlander ICE	5740	2085	6.3

- Subjects took 1.1 seconds longer on average to detect vehicles in the high ambient sound condition than in the low ambient sound condition. The main effect of ambient is statistically significant [$F (1, 47) = 35.0$; $p < 0.05$, Greenhouse-Geisser correction for sphericity].
- The mean response time is 6.5 seconds for the low ambient and 7.7 for the high ambient. These correspond to a mean time-to-vehicle-arrival of 5.5 and 4.3 seconds for the low and high ambient condition, respectively.
- Subjects detected both ICE vehicles sooner than the HEV twins. The main effect of vehicle is statistically significant [$F (2.13, 99.9) = 106.1$; $p < 0.05$, Greenhouse-Geisser correction for sphericity].
- The interaction effect of vehicle and ambient is statistically significant [$F (2.80, 131.36) = 11.93$; $p < 0.05$, Greenhouse-Geisser correction for sphericity].
- In the low ambient sound condition, the Toyota Matrix was detected 1.2 seconds sooner on average than the Toyota Prius. The Toyota Highlander ICE vehicle was detected 1.5 seconds sooner on average than the Toyota Highlander hybrid.
- In the high ambient sound condition, the Toyota Matrix was detected 2.2 seconds sooner on average than the Toyota Prius. The Toyota Highlander ICE vehicle was detected 2.1 seconds sooner on average than the Toyota Highlander ICE vehicle.

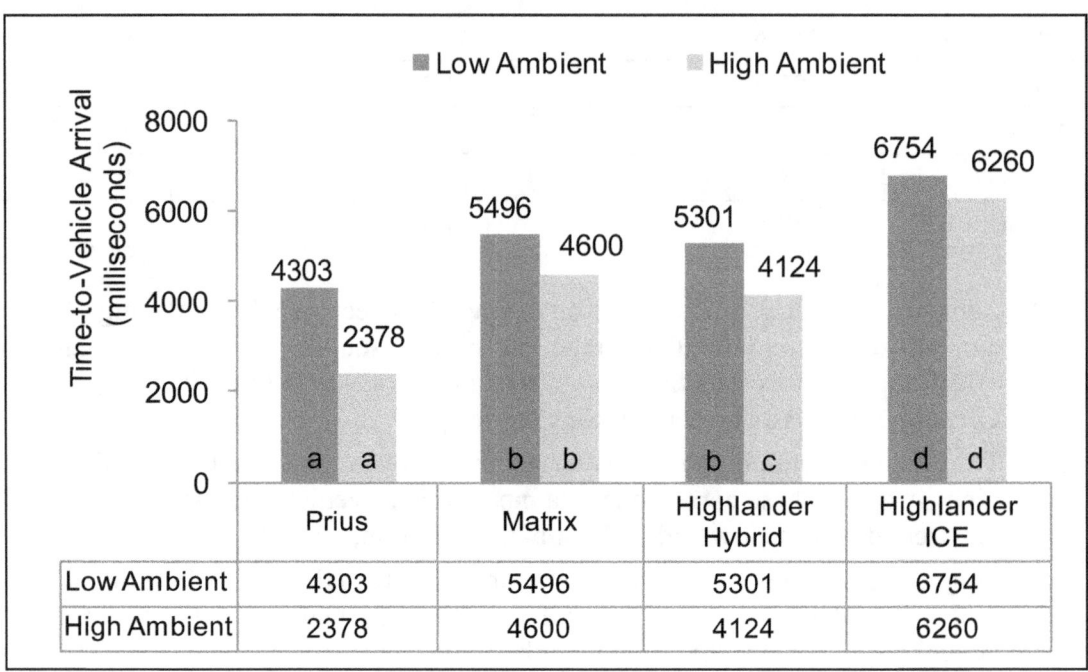

Figure 27. Mean Time-to-Vehicle-arrival: Vehicle Approaching at Low Speed

5.2.3.3 Detection Distance

The detection distance is the distance to the vehicle at the time it was detected. It is calculated by multiplying time-to-vehicle-arrival by vehicle speed (expressed in feet per second). Mean detection distances for a vehicle approaching at 6 mph in low and high ambient sound conditions are shown in Table 17 and Table 18, respectively. The stopping sight distance for a vehicle approaching at a 6 mph constant is 25.5 ft (assuming break reaction time of 2.5 s and a constant deceleration rate of 11.2 ft/s^2) (AASHTO, 2004). Thus, for the test conditions presented in this study, pedestrians detected the vehicle with enough time to avoid a potential conflict. One exception is for the Toyota Prius in the high ambient condition, where the detection distance is shorter than the vehicle stopping sight distance.

Table 17. Mean Detection Distance for Vehicle Approaching at a Constant Low Speed (Low Ambient Sound)

Vehicle	Response Time(s)	Time-to-Vehicle-arrival(s)	Detection Distance (ft)
Prius	7.7	4.3	37.9
Matrix	6.5	5.5	48.4
Highlander Hybrid	6.7	5.3	46.6
Highlander ICE	5.2	6.8	59.4

Table 18. Mean Detection Distance for Vehicle Approaching at a Constant Low Speed (High Ambient Sound)

Vehicle	Response Time(s)	Time-to-Vehicle-arrival(s)	Detection Distance (ft)
Prius	9.6	2.4	20.9
Matrix	7.4	4.6	40.5
Highlander Hybrid	7.9	4.1	36.3
Highlander ICE	5.7	6.3	55.1

5.3 Discussion

Time-to-vehicle-arrival is the difference between the time the vehicle arrives at the microphone line/pedestrian location and the time at which a subject detects its approach. Times-to-vehicle-arrival that are too short could have negative impacts on pedestrian safety. Three scenarios were evaluated: vehicle backing out; vehicle in parallel and slowing; and vehicle at approaching at a constant low speed. Average times-to-vehicle-arrival are shown in Table 19:

Table 19 Average Times to Vehicle-Arrival (seconds)

Vehicle Maneuver	Ambient Sound Level			
	Low		High	
	HEVs	ICE Vehicles	HEVs	ICE Vehicles
Backing out (5 mph)	3.7	5.2	2.0	3.5
Slowing from 20 to 10 mph	2.5	1.3	2.3	1.1
Approaching at 6 mph	4.8	6.2	3.3	5.5

The results of the human subject studies show that response time for each vehicle maneuver depends on ambient sound level and vehicle type. Overall, vehicles are detected sooner in the low ambient condition. ICE vehicles tested are detected sooner than their HEV twins except for the vehicle slowing scenario where HEVs were detected sooner. The trend observed in the vehicle slowing scenario (i.e., HEVs are detected sooner than their ICE vehicle twins) may be explained by the noticeable peak in the 5000 Hz one-third octave band for the Toyotas during this operation. These tones are much more visible in narrower-band plots (e.g., one-tenth octave or less).

Although the times-to-vehicle-arrival for the HEVs are small, they are usually sufficient for the pedestrian to take some evasive action or raise their white cane to enhance conspicuity. However, in discussions with subjects, the most difficult scenarios were those in which HEVs appeared unexpectedly (as in backing out of a driveway or parking space) as opposed to a crosswalk at a street corner. The experiments in this study mimic the situation in which a blind pedestrian knows there is a high probability of hearing a vehicle within a few seconds and can devote full attention to listening for it. Unfortunately, there is no cost-effective experimental technique to reproduce the situations in which there are long intervals between the occurrences of vehicle sounds, where pedestrians may be distracted by other matters. It is reasonable to expect that times to-vehicle-arrival for very quiet vehicles would be even shorter than the time-to-vehicle-arrival measured in this study when pedestrians are distracted.

6. EXAMINATION OF POTENTIAL COUNTERMEASURES

Of the more than 20 million Americans with some degree of vision impairment, roughly 1.3 million are legally blind. Of these, about 100,000 are "independent travelers" who use white canes in coordination with their other senses for orientation and mobility. This group is one of the most at risk of conflicts with quiet vehicles.

When independent travelers arrive at a point of possible conflict with vehicular traffic they seek two types of information:

- When it is safe to cross; and
- After initiating a crossing into the path of oncoming vehicles by extending a white cane, confirmation that vehicles are in fact slowing to a stop.

Auditory cues are the primary means of obtaining both types of information for independent travelers. These cues are generated principally by tire noise where vehicles are traveling at speeds above 20 mph. At lower speeds, internal-combustion-engine noise is the dominant source for cues. Its absence in HEVs results in the loss of both types of needed information.

6.1 Countermeasure Alternatives

Various countermeasure concepts to compensate to some extent for the loss of auditory cues have been suggested. These are summarized in Table 20. From this table, it is evident that most of the concepts have serious shortcomings:

- Infrastructure-based concepts tend to have very long implementation times. Because they entail significant capital outlays at every crossing where they are installed, it is doubtful that they will ever be deployed at more than a small fraction of all possible places where vehicle-pedestrian conflicts can occur.
- Orientation and mobility training programs for independent travelers and service animals already include quiet cars. Guide dogs have always been trained to recognize approaching vehicles by sight rather than sound, so there is no potential for further safety gain through that approach.
- Environmental regulations to lower overall urban noise levels face very long implementation times.
- There is strong objection from the blind community regarding the use of pedestrian-carried electronic warning systems.

Table 20. Pedestrian Safety Countermeasures

Category	Countermeasure	Description	Potential Benefits	Shortcomings/ Challenges	Development Status
Infrastructure-based	Accessible pedestrian signals	Device that communicates information about pedestrian timing in non-visual format; such as audible tones, verbal messages, and/or vibrating surfaces.	Allow more accurate judgments of the onset of the walk interval. Reduce the number of crossings begun during the "Don't Walk Interval." Reduce pedestrian delay.	Disagreement among blind people on the need for, and effectiveness of, audible pedestrian signals. Applicable only at signalized intersections, which constitute only a small fraction of possible points of danger. Installed at only a tiny fraction of signalized intersections. Inform blind pedestrian he/she has right-of-way, but no feedback as to whether vehicle is actually slowing or not.	Available and installed at a tiny fraction of intersections.
	Automatic pedestrian detection systems for uncontrolled approaches.	Uncontrolled crosswalks are fitted with automated detection devices that activate flashing beacons, in-pavement raised markers with LED strobe lights, or other active warnings.	Alert drivers when pedestrians are present.	Detection accuracy to reduce the number of false alarms and missed calls.	Prototyped
	Rumble strips/sound strips	Located near the crosswalk; generates noise as vehicles approach.	Improved probability of vehicle detection.	May cause noise pollution and community opposition.	Available

Chapter 6: Examination of Potential Countermeasures

Category	Countermeasure	Description	Potential Benefits	Shortcomings/ Challenges	Development Status
Education & Enforcement	Orientation and mobility training for blind pedestrians and guide dogs	Guide dogs are already trained to rely on vision, not sound.	If the dog senses danger, it can ignore a command to cross the street, or alert its owner to possible impediments.	Limited to a small fraction of the blind community who use dogs.	Implemented
Environmental Regulation	Initiatives to reduce ambient noise	Lower ambient sound levels would enhance detectability of approaching vehicles.	Improved detectability for all vehicles.	Difficult to reduce ambient levels due to non-vehicular sources (e.g., construction bldg ventilation, vehicles, animals, wind, etc).	Proposed
Vehicle-based	Artificial engine sound	Engine and exhaust noises projected through front and rear speakers that simulate those of an ICE engine, including RPMs, starting noises, etc.	Provides same minimum amount of information as ICE vehicles.	May cause noise pollution and community opposition; May increase vehicle cost. Concerns about driver acceptance	Prototyped
Vehicle-Pedestrian Communication	Proximity warning system	Battery-operated transmitter that would be carried by the pedestrian and a receiver mounted on the vehicle. Warning emitted to both pedestrian and driver.	Provides information to both the driver and pedestrian about a potential conflict.	May require integration with other in-vehicle systems. Concerns about driver and pedestrian acceptance. Devices proposed thus far do not provide cues about vehicle speed and rate of speed change.	Prototyped

Category	Countermeasure	Description	Potential Benefits	Shortcomings/ Challenges	Development Status
Pedestrian-based	Electronic travel aids	Handheld or attached to the cane. Provide tactile or audio output to inform pedestrians about their surroundings and nearby vehicles.	Provides information for avoidance of obstacles and/or vehicles, Detects distance and direction of obstacles and/or vehicles.	Range of detection. User acceptance. Battery replacement. May require additional training. Cost.	Available/ conceptualized

Because of the lack of effectiveness and/or long implementation times of other approaches, the countermeasures identified for possible further consideration in this study are the devices that generate synthetic engine noises whenever a vehicle is operating at low speed, and those that generate other types of audible alert signals (e.g., beeping) in response to a wireless signal from a transmitter carried by pedestrians.

6.1.1 Vehicle-based Audible Alert Signals

The concern that quiet electric vehicles pose a hazard for pedestrians was recognized by the vehicle manufacturers long before these vehicles gained significant market share. The development of low-cost, digital-signal-processing (DSP) integrated circuits in the 1990s made it economically feasible to generate synthetic sounds that could accurately mimic the sounds of ICE vehicles. In 1994, Honda Motor Company applied for a U.S. patent for a "Simulated Sound Generator for Electric Vehicles." Patent number 5,635,903 was awarded to Honda on June 3, 1997.

The essential elements of an effective ICE vehicle synthetic noise warning system include:
- Sensors to detect when the vehicle is started, accelerator position, and rotational rate sensor(s) on the motor and/or drive shaft, provides data for both speed and direction of operation;
- Programming/user input options to select the types of sounds that will be generated in response to the sensor inputs;
- A sound generator DSP chip;
- An ambient noise sensor (microphone) that provides input to a circuit that adjusts the output noise level to a value appropriate for particular situation; and
- Amplifiers and loudspeakers facing both forward and rearward.

Nissan Motor Company has demonstrated a sound generation system for its forthcoming electric vehicles, and has been conducting experiments to evaluate public acceptance and recognition for various sounds. In a press release dated September 18, 2009, Nissan engineer Toshiyuki Tabata said that the company favors a futuristic sound for its EVs more akin to a turbine than a piston engine. However, the issue of how these cars should sound is under review by the Japan Automobile Manufacturers Association. Guidelines from the Japanese government are anticipated in 2010.

The most widely publicized effort to provide synthetic engine noise for quiet vehicles is from Lotus Engineering in partnership with Harman International. Lotus Engineers, which have been working on active noise-cancellation (ANC) for more than 20 years, realized that the same devices used for ANC could also generate synthetic engine noise to warn pedestrians of quiet vehicles. Lotus named their prototype alert system for HEVs, *Safe and Sound* (now called "HALOsonic External Sound Synthesis" (Lotus Engineering, 2009)). These units have been designed for installation in a Prius, and can be user-controlled to sound like any of six different ICE vehicles or two futuristic vehicles.

The engineering of a sound generation system for a quiet vehicle is relatively simple, especially since the signals for input variables are already present on the Controller Area Network bus (CANbus)—the system through which the various microprocessors in all modern motor vehicles communicate. Various entrepreneurs (apparently students) have issued press releases or created websites describing sound generating systems for quiet vehicles. Some of these developers emphasize the ability of their devices to make a quiet car sound like something completely different—an exotic sports car, a motorcycle, or something out of science fiction.

6.1.2 Systems Requiring Vehicle-Pedestrian Communications

To avoid having quiet vehicles generate additional sound at all times when they are operating at low speeds, some developers have proposed systems in which a quiet vehicle emits additional sound (usually beeping) only when it receives a radio-frequency signal from a transmitter carried by someone who wants to be alerted of the presence of quiet vehicles. These systems can easily be designed to have very short range, e.g., 200 ft or less. The transmitters could operate continuously, causing any equipped vehicle to emit its warning sound whenever a transmitter is nearby, or they could have push-button actuation so that a blind pedestrian would receive warnings only when desired, as at a street crossing.

Creative Performance Products, Inc. has developed a prototype of such a device, and named it "PASS" (for Proximity Alarm Safety System). In its current implementation, this device is a simple beeper, the operation of which can be viewed and heard at the following URL:
http://www.cprracing.com/new_page_1.htm

As a simple beeper, it provides information about the direction of an approaching vehicle, but does not provide cues about its rate of change of speed. However, this capability could be added using the same kind of DSP-based sound generator as the other synthetic engine noise generators.

6.2 Advantages and Disadvantages of Potential Countermeasures

The following criteria were used when reviewing potential countermeasures: detection range, alert time, types of information provided (direction, vehicle speed), acceptability, and barriers to implementation. Although infrastructure enhancements and orientation and mobility training can generally reduce accident risks for blind pedestrians, they cannot directly address the problem that HEVs are harder to hear or inaudible in many situations in which conventional vehicles are clearly audible. In theory, radio-frequency-based systems could provide such alerts. However, to date none have been demonstrated that combine instantaneous warnings with directional and speed cues.

At present, only countermeasures that cause quiet vehicles to emit additional sound come close to meeting the requirements of blind pedestrians. Within this class of countermeasures, there is a fundamental distinction between systems that emit synthetic engine noise at all times when the vehicle is operating at low speeds, and those that emit noise only when triggered by a transmitter carried by blind pedestrians. The former eliminate the need for blind pedestrians to carry special transmitters, and also warn other pedestrians, cyclists and animals of the approach of quiet vehicles, while the latter minimize community noise impact.

6.3 Discussion

In an isolated, single-vehicle detection experiment, such as described in this report, any sound emitted by an approaching vehicle can serve as a warning to pedestrians. Even very faint sounds, e.g., the 5 kHz tone emitted by the electronics in regenerative braking mode, can provide a useful warning in some ambient-noise conditions. However, in real-world situations with multiple vehicles traveling in various directions at numerous locations around a pedestrian, it is intuitively obvious that some sounds will provide more effective alert than others. The following list of considerations for the design of a sound generation system for HEVs is presented:

1. Higher frequency sounds (above 1 kHz) are more useful for alerting than lower frequency sounds. This results from the fact that human hearing is most sensitive in the 1 to 5 kHz range. (Fletcher & Munson, 1933), as explained in the Wikipedia article at this link, http://en.wikipedia.org/wiki/Fletcher%E2%80%93Munson_curves.

   Furthermore, at shorter wavelengths intra-aural phase differences become large, and complex interactions between the pinna (outer ear) and the incident sound waves occur, both of which provide directional cues. This suggests that HEVs should emit higher frequency sounds that are approximately as loud as those emitted by ICE vehicles.

2. Conversely, lower frequency sounds are less useful for short-range acoustic alerts. Sound generators for HEVs can be designed to have less output in this range than conventional vehicles. This will lower their overall community

noise impact compared with ICE vehicles, without compromising the usefulness of their alert. (Fletcher & Munson, 1933).

3. Complex sounds that extend over a considerable range of frequencies—typically an octave or more—are more effective for alert than single-tone sounds. A broad range of frequencies produces a more complex set of interactions with the ears of the listener, which in turn yields better localization of the sound source.

4. Research on recognition of auditory warnings has shown that those warnings based on the actual sound made by the object of the warning are more quickly recognized than abstract sounds (tones, chimes, etc.) or simple computer-generated sounds (e.g., new mail signal in Windows) (Leung, Smith, Parker, & Martin ,1997; Stevens, Perry, Wiggins, & Howell, 2006).

5. Warning sounds should be designed to attract attention without generating a startle effect. Dissonances and high levels of harmonic distortion should be avoided because they tend to annoy listeners (Federal Highway Administration, 2004).

6. When ambient noise is low, the amount of sound that must be emitted by an HEV to provide sufficient warning to pedestrians is also lower. By adjusting its sound output to an appropriate level for a particular environment, an HEV can be more detectable for blind pedestrians, while having less overall community noise impact.

7. Ideally, the sound emitted by HEVs should mimic that of an ICE vehicle operating at the same speed and rate of change of speed to make recognition of the alert sounds intuitive for all pedestrians who can hear.

8. The characteristic sound of an ICE vehicle being started is often the first cue that a blind pedestrian has regarding the presence of a new threat when walking through a parking area. Groups representing people who are blind have indicated preference that HEVs also mimic this sound.

9. At speeds above 20 mph, tire noise becomes dominant and the sound output level of HEVs and ICE vehicles are essentially the same. There is no need for the sound generation system to operate at speeds above that at which tire-noise dominates. Further experiments will refine estimates of the speed at which the sound generator can be automatically switched off.

These insights suggest that the sounds of HEVs should be recognizable as a vehicle to be most helpful to pedestrians. Nancy Gioia, director of global electrification at Ford, said in a *New York Times* interview published October 15, 2009:

"….if cars and trucks emit personalized noise, the sound—from Iron Butterfly singing "In-A-Gadda-da-Vida," to a veritable symphony of rings and tones—would get lost in the general din. It will get lost and it won't meet the objective of being a sound that lets you know a car is coming."

There is a potential conflict between the need for standardization of vehicle sounds to enhance recognition versus possible end-user preference for unique, personalized sounds. Even if vehicle manufacturers offer only sound generators that conform to norms for ICE vehicles, it is also possible that after-market vendors may attempt to sell non-conforming products. Furthermore, because of the relative ease with which DSP chips can be reprogrammed, hacking of sound generators to produce personalized sounds is a distinct possibility.

7. SUMMARY OF FINDINGS AND CONSIDERATIONS FOR FURTHER RESEARCH

7.1 Findings

This research provides information to better understand the safety risk to blind pedestrians associated with the acoustic profiles of quieter cars in various maneuvers and ambient sound conditions. Pedestrians use acoustic cues to monitor vehicles, support crossing strategies, and avoid conflicts. The research has identified the information that blind pedestrians depend on for safe travel. Pedestrians assess risks from vehicles approaching at a constant speed, vehicles turning, and vehicles backing out into their path. Pedestrians have to determine the presence of a vehicle of interest, its relative position, and direction of travel, as well as its rate of acceleration to judge how fast the vehicle is moving or how soon it may reach their position. Information comes from vehicles that are accelerating, decelerating, or idling. Critical safety scenarios were identified, which include vehicle slowing (as if to turn right from the parallel street); vehicle moving in reverse; and vehicle approaching at a constant low speed. These high risk scenarios occur in proximity to driveways, controlled intersections, uncontrolled approaches, and when vehicles turn into or back up into the pedestrian's path.

The SAE test procedure for acoustic measurement of vehicles was reviewed and adapted for use in the test plan to measure HEV and ICE vehicle acoustic parameters. This test plan was implemented in recording the vehicle sounds emitted by HEVs and ICE vehicles operated under conditions simulating critical safety scenarios. The Volpe Center team combined the audio recordings with ambient recordings for use in human subject testing. There is a noticeable difference in vehicle sound by vehicle type, which is consistent with previous studies. The difference in sound also varies by vehicle maneuver or operating condition.

The following findings were obtained for the selection of vehicles tested:

- The overall sound levels for HEVs traveling in reverse at 5 mph are 7 to 10 dB(A) lower than for their ICE vehicle twins (ranged from 44.2 to 48.5 dB(A) for the HEVs tested and 51.3 to 58.2 dB(A) for the ICE vehicles).
- The overall sound levels for HEVs traveling in parallel and slowing (from 20 mph to 10 mph) are 1.2 to 2.4 dB(A) lower than for their ICE vehicle twins (ranged from 53.0 to 56.6 dB(A) for HEVs and from 54.2 to 55.4 dB(A) for ICE vehicles).
- The overall sound levels for HEVs approaching at 6 mph were 2 to 8 dB(A) lower for the HEVs than for their ICE vehicle twins (ranged from 44.7 to 53.2 dB(A) for HEVs and from 52.0 to 55.5 dB(A) for ICE vehicles).
- Overall sound levels at 10 mph were 0.6 to 2.4 dB(A) lower for the HEVs than for their ICE vehicle twins (HEVs ranged from 44.7 to 53.2 dB(A) and from 52.0 to 55.5 dB(A) for ICE vehicles).
- The sound levels for HEVs and ICE vehicles converge at higher speeds. The speed at which vehicles converge varies between the three sets of vehicles tested (ranged from 10 to about 20 mph).

- The overall sound levels for HEVs are 0.1 to 1.9 dB(A) lower than for their ICE vehicle twins when accelerating from a stop.
- The sound levels for the Toyota hybrids when stationary were too low to be measured under the ambient condition present.
- The sound levels for the Honda Civic Hybrid and for the other three ICE vehicles ranges from 46.0 to 47.9 dB(A) when stationary.
- Considering the one-third octave band spectrum, there is a trend for HEVs to have less high frequency content relative to the overall sound level. The Toyota Hybrids are an exception to this trend with a notable peak in the 5 kHz one-third octave band sound level when slowing or braking.

In addition, the research examined how the acoustic characteristics of vehicles and ambient sound affect blind pedestrians' detection of vehicles. Laboratory studies were carried out using subjects who are blind to determine whether and how quickly they can detect vehicles approaching in three operating conditions: vehicle backing out; vehicle in parallel and slowing; and vehicle approaching at 6 mph. The measures considered include: missed detection; response time; time-to-vehicle-arrival; and detection distance. The numbers of subjects that never detected a vehicle in a given vehicle-ambient sound condition vary for the three operating conditions. Five subjects (10.4%) never detected one or more HEVs backing out; eight subjects (16.7%) never detected slowing vehicles (most frequently the Matrix); and two subjects (4.2%) never detected a Toyota Prius approaching at 6 mph.

Time-to-vehicle-arrival was significantly affected by ambient sound. Overall, subjects detected vehicles later (i.e., closer to vehicle arrival at the pedestrian position) in the high ambient sound condition than in the low ambient sound condition. The average times-to-vehicle-arrival for the low ambient condition are: 4.4 s for vehicles backing out; 1.9 s for vehicles moving in parallel and slowing; and 5.5 s for vehicles approaching at a constant speed. The corresponding values for the high ambient condition are: 2.7 s for vehicles backing out; 1.7 s for vehicles moving in parallel and slowing; and 4.4 s for vehicles approaching at a constant speed.

Time-to-vehicle-arrival was also significantly affected by vehicle type. In the low ambient condition, subjects detected both ICE vehicles backing out sooner than their HEV twins. On average, the Toyota Matrix was detected 1.2 seconds sooner than the Toyota Prius and the Toyota Highlander ICE vehicle was detected 1.9 seconds sooner than the Toyota Highlander hybrid. In the high ambient condition, subjects detected both ICE vehicles sooner than their HEV twins. On average, the Toyota Matrix was detected 1.1 seconds sooner than the Toyota Prius and the Toyota Highlander ICE vehicle was detected 1.9 seconds sooner than the Toyota Highlander hybrid.

Toyota HEVs were detected sooner than their ICE vehicle twins when slowing from 20 mph to 10 mph. On average, in the low ambient condition, the Toyota Prius was detected 0.9 seconds sooner than the Toyota Matrix and the Toyota Highlander hybrid was detected 1.5 seconds sooner than the Toyota Highlander ICE vehicle. On average, in the high ambient condition the Toyota Prius was detected 1.1 seconds sooner than the Toyota Matrix and the Toyota Highlander hybrid was detected 1.4 seconds sooner than the Toyota Highlander ICE vehicle. For the test conditions presented in this study, pedestrians detected HEVs 35 to 55.1 ft before vehicle arrival and detected ICE vehicles 15 to 23.7ft before vehicle arrival. The earlier detection of the Toyota HEV is believed to have resulted from a 5 kHz tone being emitted from the electronics when these vehicles were in regenerative braking mode. Not all HEVs emit this tone.

Subjects detected both Toyota ICE vehicles sooner than their hybrid twins when approaching at a constant speed of 6 mph. On average, in the low ambient sound condition, the Toyota Matrix was detected 1.2 seconds sooner than the Toyota Prius and the Toyota Highlander ICE was detected 1.5 seconds sooner than the Toyota Highlander hybrid. On average, in the high ambient sound condition, the Toyota Matrix was detected 2.2 seconds sooner than the Toyota Prius and the Toyota Highlander ICE was detected 2.2 seconds sooner than the Toyota Highlander ICE. In the low ambient condition, pedestrians detected the Toyotas HEVs 37.9 to 46.6 ft before vehicle arrival and detected ICE vehicles 48.4 to 59.4 ft before vehicle arrival. In the high ambient condition, subjects detected the Toyota HEVs 20.9 to 36.3 ft before vehicle arrival, and detected the Toyota ICE vehicles 40.5 to 55.1 ft before vehicle arrival. The stopping sight distance for a vehicle approaching at a 6 mph constant is 25.5 ft (assuming a brake reaction time of 2.5 s and a constant deceleration rate of 11.2 ft/s^2). Therefore, in the high ambient condition the Prius may not have been detected within an adequate amount of time to avoid a collision.

7.2 Considerations for Further Research

Countermeasures that have the potential to reduce the safety risks to blind pedestrians were discussed in Chapter 6. Infrastructure-based countermeasures, such as accessible pedestrian signals (which provide an indication of the walking interval) and traffic calming applications (which reduce vehicle speed and improve access), are available and can reduce pedestrian-vehicle conflicts. Concurrent vehicle maneuvers (e.g., right turn on green and permissive left turns); vehicles running red light signals; and vehicles moving at low speeds (e.g., entering/ leaving driveways) are still a concern among pedestrians. In all these situations it is essential that the pedestrian detect the presence, direction, and intended maneuver of conflicting vehicles. Auditory cues used for travel are generated principally by tire noise when vehicles are traveling at speeds above 20 mph. Detection of HEVs operated at low speeds and in reverse is difficult, in particular when the ambient level is moderate to relatively high. Some ICE vehicles are also difficult to detect due to masking. Environmental regulations to lower overall urban noise levels face long implementation times. Active and passive pedestrian safety systems are currently available. Active safety systems provide information to the driver whereas passive safety systems can reduce the severity of an injury once the collision has occurred. However, neither of these two systems provides information to pedestrians. The countermeasures identified for possible further consideration in this study include devices that generate synthetic engine noises whenever a vehicle is operating at low speed, and those that generate other types of audible warnings (for pedestrians) in response to a wireless signal from a transmitter carried by pedestrians. These are preferred due to the lack of effectiveness of other systems to provide information to pedestrians and/or long implementation times of other approaches.

In an isolated, single-vehicle detection experiment, such as that described in this report, almost any sound emitted by an approaching vehicle can serve to alert pedestrians. However, certain sounds may also have detrimental effects. In more complex situations, with multiple vehicles traveling in various directions at numerous locations around a pedestrian, it is obvious that some sounds will provide more effective alert than others. Such cases must be considered in future evaluations. General considerations for the design of a sound generation system were discussed in Chapter 6. In particular, higher frequency sounds (above 1 kHz) appear to be more useful for alert than lower frequency sounds. Conversely, lower frequency sounds appear to be less useful for short-range acoustic alert. Sound generators for HEVs can be designed to have less output in this range than conventional vehicles. This will lower their overall community noise impact compared with ICE vehicles, without compromising the usefulness of their warnings. When ambient noise is low, the amount of sound that must be emitted by an HEV to provide sufficient warning to pedestrians is also lower. An HEV can be more detectable for blind pedestrians and have less of an overall community noise impact, by adjusting its sound output to an appropriate level for a particular environment. Pedestrians who are blind express a preference that the sound emitted by HEVs should mimic the sound of an ICE vehicle operating at the same speed and rate of change of speed. This makes recognition of the warning sounds intuitive for all pedestrians. The characteristic sound of an ICE vehicle being started is often the first cue that a blind pedestrian has regarding the presence of a new threat when walking through a parking area. It is desirable that HEVs also mimic this sound. At speeds above 20 mph, tire noise becomes dominant, and there is no need for any extra sound emission from HEVs.

8. BIBLIOGRAPHY

AASHTO. (2004). Guide for the Planning, Design, and Operation of Pedestrian Facilities. Washington, DC: American Association of State Highway and Transportation Officials, Available at: https://bookstore.transportation.org/category_item.aspx?id=DS.

Barlow, J. M., Bentzen, B. L., & Bond, T. (2005). Blind Pedestrians and the Changing Technology and Geometry of Signalized Intersections: Safety, Orientation, and Independence. *Journal of Visual Impairment & Blindness.* American Foundation for the Blind. Vol. 99, No. 10.

Barnecutt, P., & Pfeffer, K. (1998). Auditory Perception of Relative Distance of Traffic Sounds. *Current Psychology: Developmental, Learning, Personality, Social.* 17, 93-101.

Blash, B. B., Wiener, W. R., & Welsh, R. L.. (1997). Foundations of Orientation and Mobility. Second Edition. Sewickley, PA: American Foundation for the Blind Press.

Butler, R. A., Levy, E. T., & Neff, W. D. (1980). Apparent Distances of Sounds Recorded in Echoic and Anechoic Chambers. *Journal of Experimental Psychology: Human Perception and Performance.* Vol. 6, 745-750.

Campbell, J. L., Richman, J. B., Carney, C., & Lee, J. D.. (2004, September). In-Vehicle Display Icons and Other Information Display Elements, Volume I: Guidelines. FHWA-RD-03-065. Washington, DC: Federal Highway Administration.

Carroll Center for the Blind. http://www.carroll.org/.

Coleman, P. D. (1963). An Analysis of Cues to Auditory Depth Perception in Free Space, *Psychological Bulletin*, 60, 302-15.

Coleman, P. D. (1968). Dual Role of Frequency Spectrum in Determination of Auditory Distance. *Journal of Acoustical Society of America*, 44, 631-632.

Eye Diseases Prevalence Research Group. (2004). Causes and Prevalence of Visual Impairment Among Adults in the United States. *Archives of Ophthalmology.* Volume 122. pp. 477-485.

Fletcher, H., & Munson, W. A. (1933). Loudness, Its Definition, Measurement, and Calculation. *Journal of the Acoustic Society of America.* Volume 5, Issue 1. pp. 82-108.

Geruschat, D. R., & Hassan, S. E. (2005). Driver Behavior in Yielding to Sighted and Blind Pedestrians at Roundabouts. *Journal of Visual Impairment & Blindness.* May 2005. Pp. 286-302.

Guth, D., Ashmead, D., Long, R., Wall, R., & Ponchillia, P. (2005). Blind and Sighted Pedestrians' Judgments of Gaps in Traffic at Roundabouts. *Human Factors.* Volume 47, No 2.

Hanna, R.. (2009, September). Incidence of Pedestrians and Bicyclist Crashes by Hybrid Electric Passenger Vehicles: National Center for Statistical Analysis Technical Report. DOT HS 811 204 Washington, DC: National Highway Traffic Safety Administration. Available at: http://www-nrd.nhtsa.dot.gov/Pubs/811204.PDF.

Ito, K. (1997). Auditory Interceptive Timing and Familiarity with Acoustic Environment. *Studies in Perception and Action IV*. pp. 83-87.

Japanese Automobile Standards Internationalization Centre. (2009). A Study on Approach Warning Systems for Hybrid Vehicle in Motor Mode. Presented at the 49th World Forum for Harmonization of Vehicle Regulation (WP.29) GRB Working Group on Noise. February 16-18, 2009. Document Number: GRB-49-10. Available at: www.unece.org/trans/doc/2009/wp29grb/ECE-TRANS-WP29-GRB-49-inf10e.pdf.

Leung, Y. K., Smith, S, Parker, S., & Martin, R. (1997). Learning and Retention of Auditory Warnings. International Conference on Auditory Display. Palo Alto, CA. 1997. Available at: http://www.icad.org/node/2924.

Lotus Engineering. (2009). HALOsonic Noise Management Solutions: External Electronic Sound Synthesis (EESS). Norfolk, UK: Lotus Engineering UK. Available at: http://www.grouplotus.com/engineering/home.html.

National Federation of the Blind. (2008). Resolution 2008-02 Regarding Momentum Toward Solving the Quiet Cars Crisis. *The Braille Monitor*. August/September 2008. Available at: http://www.nfb.org/images/nfb/Publications/bm/bm08/bm0808/bm080812.htm.

National Federation of the Blind. (2008). The Danger Posed by Silent Vehicles. Remarks by Marc Maurer. Work Forum for Harmonization of Vehicle Regulations (United Nations Working Party 29). February 20, 2008, Geneva, Switzerland. Baltimore, MD: National Federation of the Blind. .

National Federation of the Blind. (2008). Presentation documented in the Transcript of the Quiet Cars Public Meeting on June 23, 2008. Docket ID NHTSA-2008-0108-0023. *Statement of Problem*. pp. 18-24. Washington, DC: National Highway Traffic Safety Administration. Available at: http://www.regulations.gov/fdmspublic/component/main?main=DocumentDetail&o=09000064806e5c9c.

NHTSA, Quieter Cars and the Safety of Blind Pedestrians: A Research Plan. April 2009. NHTSA Quiet Cars - Notice and Request for Comments (2008) Docket ID NHTSA-2008-0108-0025. Washington, DC: National Highway Traffic Safety Administration.

NHTSA Quiet Cars - Notice and Request for Comments (2008) Docket ID NHTSA-2008-0108-001. Washington, DC: National Highway Traffic Safety Administration.

Perkins School for the Blind. http://www.perkins.org/.

Rosenblum, L. (2008). *Sound Measurement and Mobility*. pp. 53 -65. Transcript of the Quiet Cars Public Meeting on June 23, 2008. Docket ID NHTSA-2008-0108-0023. Washington, DC: National Highway Traffic Safety Administration

SAE. (2009, February). Surface Vehicle Standard. Measurement of Minimum Noise Emitted by Road Vehicles. J2889-1 Proposed Draft. Warrendale, PA: Society of Automotive Engineers.

Sauerburger, D. (2005). Street Crossings: Analyzing Risk, Developing Strategies, and Making Decisions. *Journal of Visual Impairment & Blindness*. Vol. 99, No. 10. pp 659.

Schroeder, B. J., Rouphail, N. M., & Wall Emerson, R. (2006). Exploratory Analysis of Crossing Difficulties for Blind and Sighted Pedestrians at Channelized Turn Lanes. *Transportation Research Record.* No. 1956. pp. 94-102.

Stevens, C., Perry, N., Wiggins, M., & Howell, C. (2006). Design and Evaluation of Auditory Icons as Informative Warning Signals. Canberra, Australia: Australian Transportation Safety Bureau. Available at: http://www.atsb.gov.au/publications/2006/grant_B20050120_001.aspx.

Transportation Research Center. http://www.trcpg.com/.

Wall Emerson, R., & Sauerburger, D. (2008). Detecting Approaching Vehicles at Streets with No Traffic Control. *Journal of Visual Impairment & Blindness.* December 2008.

Wiener, W., & Lawson, G. (1997). The Use of Traffic Sounds to Make Street Crossings. *Journal of Visual Impairment & Blindness.* Volume 91, Issue (5). pp. 435.

Wiener, W., Naghshineh, K., Salisbury, B., & Rozema, R. (2007). The Impact of Hybrid Vehicles on Street Crossings. Re:View: Rehabilitation Education for Blindness and Visual Impairment. Volume 38, Number (2). pp 65-78. Available at: http://search.ebscohost.com/login.aspx?direct=true&db=afh&AN=25893339&site=ehost-live

APPENDIX A. ACOUSTIC DATA FOR VEHICLES

LIST OF TABLES

Table A-1. Overall Levels by Vehicle for Idle at 12 ft Microphone A-32
Table A-2. Overall Levels by Vehicle for 6 mph Constant Speed
 Passby at 12 ft Microphone A-32
Table A-3. Overall Levels by Vehicle for 10 mph Constant Speed
 Passby at 12 ft Microphone A-32
Table A-4. Overall Levels by Vehicle for 20 mph Constant Speed
 Passby at 12 ft Microphone A-33
Table A-5. Overall Levels by Vehicle for 30 mph Constant Speed
 Passby at 12 ft Microphone A-33
Table A-6. Overall Levels by Vehicle for 40 mph Constant Speed
 Passby at 12 ft Microphone A-33
Table A-7. Overall Levels by Vehicle for Acceleration at 12 ft Microphone A-34
Table A-8. Overall Levels by Vehicle for Deceleration at 12 ft Microphone A-34
Table A-9. Overall Levels by Vehicle for Reverse (5 mph Constant Speed Passby)
 at 12 ft Microphone A-34

LIST OF FIGURES

Figure A-1. Prius Time History for Idle at 12 ft Microphone A-6
Figure A-2. Prius Time History for 6 mph Constant Speed Passby at 12 ft Microphone A-6
Figure A-3. Prius Time History for 10 mph Constant Speed Passby at 12 ft Microphone A-7
Figure A-4. Prius Time History for 20 mph Constant Speed Passby at 12 ft Microphone A-7
Figure A-5. Prius Time History for 30 mph Constant Speed Passby at 12 ft Microphone A-8
Figure A-6. Prius Time History for Acceleration Passby at 12 ft Microphone A-8
Figure A-7. Prius Time History for Reverse 5 mph Constant Speed
 Passby at 12 ft Microphone A-9
Figure A-8. Prius Time History for Deceleration Passby at 12 ft Microphone A-9
Figure A-9. Prius Time History for 40 mph Constant Speed Passby at 12 ft Microphone A-10
Figure A-10. Matrix Time History for Idle at 12 ft Microphone A-10
Figure A-11. Matrix Time History for 6 mph Constant Speed Passby at 12 ft Microphone A-11
Figure A-12. Matrix Time History for 10 mph Constant Speed Passby at 12 ft Microphone .. A-11
Figure A-13. Matrix Time History for 20 mph Constant Speed Passby at 12 ft Microphone .. A-12
Figure A-14. Matrix Time History for 30 mph Constant Speed Passby at 12 ft Microphone .. A-12
Figure A-15. Matrix Time History for Acceleration Passby at 12 ft Microphone A-13

Figure A-16. Matrix Time History for Reverse 5 mph Constant Speed
  Passby at 12 ft Microphone A-13
Figure A-17. Matrix Time History for Deceleration Passby at 12 ft Microphone A-14
Figure A-18. Matrix Time History for 40 mph Constant Speed Passby at 12 ft Microphone ... A-14
Figure A-19. Honda Civic Hybrid Time History for Idle at 12 ft Microphone A-15
Figure A-20. Honda Civic Hybrid Time History for 6 mph Constant Speed
  Passby at 12 ft Microphone A-15
Figure A-21. Honda Civic Hybrid Time History for 10 mph Constant Speed
  Passby at 12 ft Microphone A-16
Figure A-22. Honda Civic Hybrid Time History for 20 mph Constant Speed
  Passby at 12 ft Microphone A-16
Figure A-23. Honda Civic Hybrid Time History for 30 mph Constant Speed
  Passby at 12 ft Microphone A-17
Figure A-24. Honda Civic Hybrid Time History for Acceleration
  Passby at 12 ft Microphone A-17
Figure A-25. Honda Civic Hybrid Time History for Reverse 5 mph Constant Speed
  Passby at 12 ft Microphone A-18
Figure A-26. Honda Civic Hybrid Time History for Deceleration
  Passby at 12 ft Microphone A-18
Figure A-27. Honda Civic ICE Time History for Idle at 12 ft Microphone................................ A-19
Figure A-28. Honda Civic ICE Time History for 6 mph Constant Speed
  Passby at 12 ft Microphone A-19
Figure A-29. Honda Civic ICE Time History for 10 mph Constant Speed
  Passby at 12 ft Microphone A-20
Figure A-30. Honda Civic ICE Time History for 20 mph Constant Speed
  Passby at 12 ft Microphone A-20
Figure A-31. Honda Civic ICE Time History for 30 mph Constant Speed
  Passby at 12 ft Microphone A-21
Figure A-32. Honda Civic ICE Time History for Acceleration
  Passby at 12 ft Microphone A-21
Figure A-33. Honda Civic ICE Time History for Reverse 5 mph Constant Speed
  Passby at 12 ft Microphone A-22
Figure A-34. Honda Civic ICE Time History for Deceleration
  Passby at 12 ft Microphone A-22
Figure A-35. Toyota Highlander Hybrid Time History for Idle at 12 ft Microphone A-23
Figure A-36. Toyota Highlander Hybrid Time History for 6 mph Constant Speed
  Passby at   12 ft Microphone A-23
Figure A-37. Toyota Highlander Hybrid Time History for 10 mph Constant Speed
  Passby at 12 ft Microphone A-24
Figure A-38. Toyota Highlander Hybrid Time History for 20 mph Constant Speed
  Passby at 12 ft Microphone A-24
Figure A-39. Toyota Highlander Hybrid Time History for 30 mph Constant Speed
  Passby at 12 ft Microphone A-25
Figure A-40. Toyota Highlander Hybrid Time History for Acceleration
  Passby at 12 ft Microphone A-25

Figure A-41. Toyota Highlander Hybrid Time History for Reverse 5 mph Constant Speed Passby at 12 ft Microphone A-26

Figure A-42. Toyota Highlander Hybrid Time History for Deceleration Passby at 12 ft Microphone A-26

Figure A-43. Toyota Highlander Hybrid Time History for 40 mph Constant Speed Passby at 12 ft Microphone A-27

Figure A-44. Toyota Highlander ICE Time History for Idle at 12 ft Microphone.....................A-27

Figure A-45. Toyota Highlander ICE Time History for 6 mph Constant Speed Passby at 12 ft Microphone A-28

Figure A-46. Toyota Highlander ICE Time History for 10 mph Constant Speed Passby at 12 ft Microphone A-28

Figure A-47. Toyota Highlander ICE Time History for 20 mph Constant Speed Passby at 12 ft Microphone A-29

Figure A-48. Toyota Highlander ICE Time History for 30 mph Constant Speed Passby at 12 ft Microphone A-29

Figure A-49. Toyota Highlander ICE Time History for Acceleration Passby at 12 ft Microphone A-30

Figure A-50. Toyota Highlander ICE Time History for Reverse 5 mph Constant Speed Passby at 12 ft Microphone A-30

Figure A-51. Toyota Highlander ICE Time History for Deceleration Passby at 12 ft Microphone A-31

Figure A-52. Toyota Highlander ICE Time History for 40 mph Constant Speed Passby at 12 ft Microphone A-31

Figure A-53. Prius One-Third Octave Band Levels for Idle at 12 ft Microphone.....................A-35

Figure A-54. Prius One-Third Octave Band Levels for 6 mph Constant Speed Passby at 12 ft Microphone A-36

Figure A-55. Prius One-Third Octave Band Levels for 10 mph Constant Speed Passby at 12 ft Microphone A-36

Figure A-56. Prius One-Third Octave Band Levels for 20 mph Constant Speed Passby at 12 ft Microphone A-37

Figure A-57. Prius One-Third Octave Band Levels for 30 mph Constant Speed Passby at 12 ft Microphone A-37

Figure A-58. Prius One-Third Octave Band Levels for Acceleration Passby at 12 ft Microphone A-38

Figure A-59. Prius One-Third Octave Band Levels for Reverse 5 mph Constant Speed Passby at 12 ft Microphone A-38

Figure A-60. Prius One-Third Octave Band Levels for Deceleration Passby at 12 ft Microphone A-39

Figure A-61. Prius One-Third Octave Band Levels for 40 mph Constant Speed Passby at 12 ft Microphone A-39

Figure A-62. Matrix One-Third Octave Band Levels for Idle at 12 ft Microphone A-40

Figure A-63. Matrix One-Third Octave Band Levels for 6 mph Constant Speed Passby at 12 ft Microphone A-40

Figure A-64. Matrix One-Third Octave Band Levels for 10 mph Constant Speed Passby at 12 ft Microphone A-41

Figure A-65. Matrix One-Third Octave Band Levels for 20 mph Constant Speed Passby at 12 ft Microphone A-41

Figure A-66. Matrix One-Third Octave Band Levels for 30 mph Constant Speed Passby at 12 ft Microphone A-42

Figure A-67. Matrix One-Third Octave Band Levels for Acceleration Passby at 12 ft Microphone A-42

Figure A-68. Matrix One-Third Octave Band Levels for Reverse 5 mph Constant Speed Passby at 12 ft Microphone A-43

Figure A-69. Matrix One-Third Octave Band Levels for Deceleration Passby at 12 ft Microphone A-43

Figure A-70. Matrix One-Third Octave Band Levels for 40 mph Constant Speed Passby at 12 ft Microphone A-44

Figure A-71. Honda Civic Hybrid One-Third Octave Band Levels for Idle at 12 ft Microphone A-44

Figure A-72. Honda Civic Hybrid One-Third Octave Band Levels for 6 mph Constant Speed Passby at 12 ft Microphone A-45

Figure A-73. Honda Civic Hybrid One-Third Octave Band Levels for 10 mph Constant Speed Passby at 12 ft Microphone A-45

Figure A-74. Honda Civic Hybrid One-Third Octave Band Levels for 20 mph Constant Speed Passby at 12 ft Microphone A-46

Figure A-75. Honda Civic Hybrid One-Third Octave Band Levels for 30 mph Constant Speed Passby at 12 ft Microphone A-46

-Figure A-76. Honda Civic Hybrid One-Third Octave Band Levels for Acceleration Passby at 12 ft Microphone A-47

Figure A-77. Honda Civic Hybrid One-Third Octave Band Levels for Reverse 5 mph Constant Speed Passby at 12 ft Microphone A-47

Figure A-78. Honda Civic Hybrid One-Third Octave Band Levels for Deceleration Passby at 12 ft Microphone A-48

Figure A-79. Honda Civic ICE One-Third Octave Band Levels for Idle at 12 ft Microphone ... A-48

Figure A-80. Honda Civic ICE One-Third Octave Band Levels for 6 mph Constant Speed Passby at 12 ft Microphone A-49

Figure A-81. Honda Civic ICE One-Third Octave Band Levels for 10 mph Constant Speed Passby at 12 ft Microphone A-49

Figure A-82. Honda Civic ICE One-Third Octave Band Levels for 20 mph Constant Speed Passby at 12 ft Microphone A-50

Figure A-83. Honda Civic ICE One-Third Octave Band Levels for 30 mph Constant Speed Passby at 12 ft Microphone A-50

Figure A-84. Honda Civic ICE One-Third Octave Band Levels for Acceleration Passby at 12 ft Microphone A-51

Figure A-85. Honda Civic ICE One-Third Octave Band Levels for Reverse 5 mph Constant Speed Passby at 12 ft Microphone A-51

Figure A-86. Honda Civic ICE One-Third Octave Band Levels for Deceleration Passby at 12 ft Microphone A-52

Figure A-87. Toyota Highlander Hybrid One-Third Octave Band Levels for Idle at 12 ft Microphone A-52

Figure A-88. Toyota Highlander Hybrid One-Third Octave Band Levels for 6 mph Constant Speed Passby at 12 ft Microphone A-53

Figure A-89. Toyota Highlander Hybrid One-Third Octave Band Levels for 10 mph Constant Speed Passby at 12 ft Microphone A-53

Figure A-90. Toyota Highlander Hybrid One-Third Octave Band Levels for 20 mph Constant Speed Passby at 12 ft Microphone A-54

Figure A-91. Toyota Highlander Hybrid One-Third Octave Band Levels for 30 mph Constant Speed Passby at 12 ft Microphone A-54

Figure A-92. Toyota Highlander Hybrid One-Third Octave Band Levels for Acceleration Passby at 12 ft Microphone A-55

Figure A-93. Toyota Highlander Hybrid One-Third Octave Band Levels for Reverse 5 mph Constant Speed Passby at 12 ft Microphone A-55

Figure A-94. Toyota Highlander Hybrid One-Third Octave Band Levels for Deceleration Passby at 12 ft Microphone A-56

Figure A-95. Toyota Highlander Hybrid One-Third Octave Band Levels for 40 mph Constant Speed Passby at 12 ft Microphone A-56

Figure A-96. Toyota Highlander ICE One-Third Octave Band Levels for Idle at 12 ft Microphone A-57

Figure A-97. Toyota Highlander ICE One-Third Octave Band Levels for 6 mph Constant Speed Passby at 12 ft Microphone A-57

Figure A-98. Toyota Highlander ICE One-Third Octave Band Levels for 10 mph Constant Speed Passby at 12 ft Microphone A-58

Figure A-99. Toyota Highlander ICE One-Third Octave Band Levels for 20 mph Constant Speed Passby at 12 ft Microphone A-58

Figure A-100. Toyota Highlander ICE One-Third Octave Band Levels for 30 mph Constant Speed Passby at 12 ft Microphone A-59

Figure A-101. Toyota Highlander ICE One-Third Octave Band Levels for Acceleration Passby at 12 ft Microphone A-59

Figure A-102. Toyota Highlander ICE One-Third Octave Band Levels for Reverse 5 mph Constant Speed Passby at 12 ft Microphone A-60

Figure A-103. Toyota Highlander ICE One-Third Octave Band Levels for Deceleration Passby at 12 ft Microphone A-60

Figure A-104. Toyota Highlander ICE One-Third Octave Band Levels for 40 mph Constant Speed Passby at 12 ft Microphone A-61

A.1 Time History (12-ft Microphone)

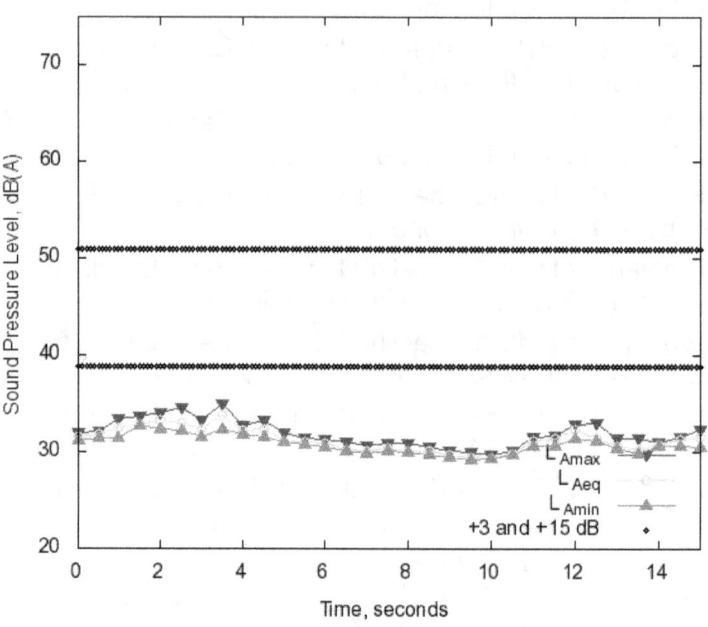

Figure A-1. Prius Time History for Idle at 12 ft Microphone

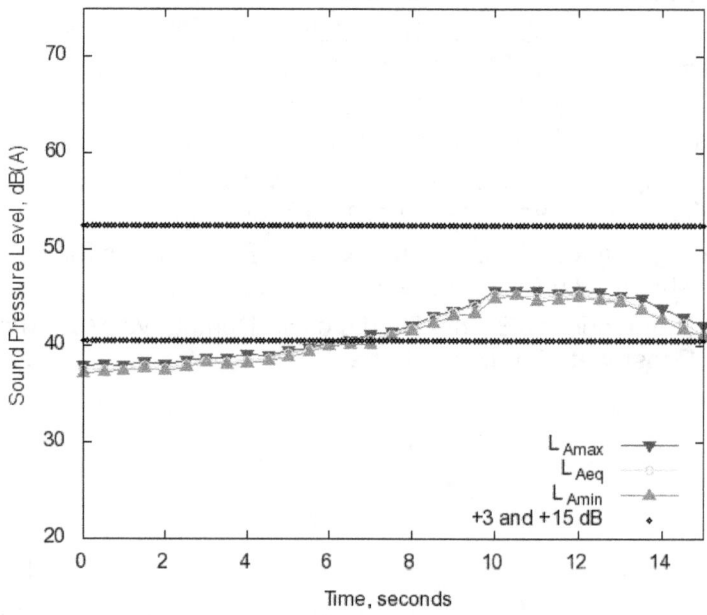

Figure A-2. Prius Time History for 6 mph Constant Speed Passby at 12 ft Microphone

Appendix A: Acoustic Data for Vehicles

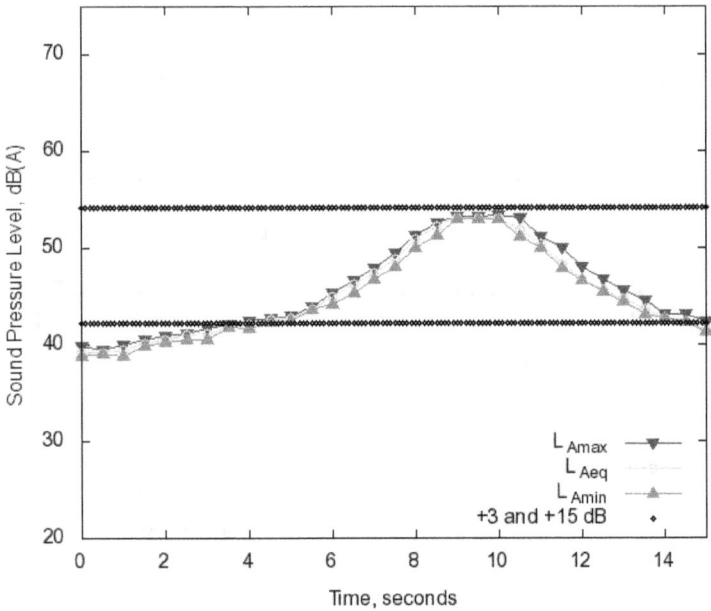

Figure A-3. Prius Time History for 10 mph Constant Speed Passby at 12 ft Microphone

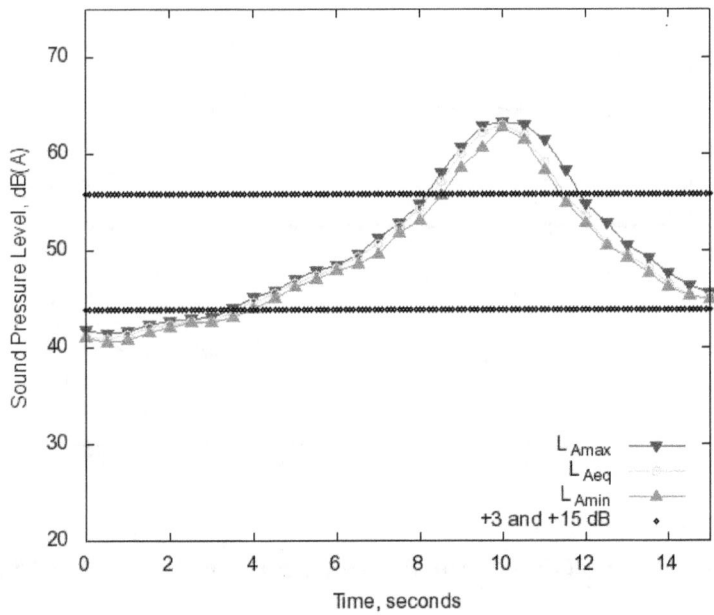

Figure A-4. Prius Time History for 20 mph Constant Speed Passby at 12 ft Microphone

Appendix A: Acoustic Data for Vehicles

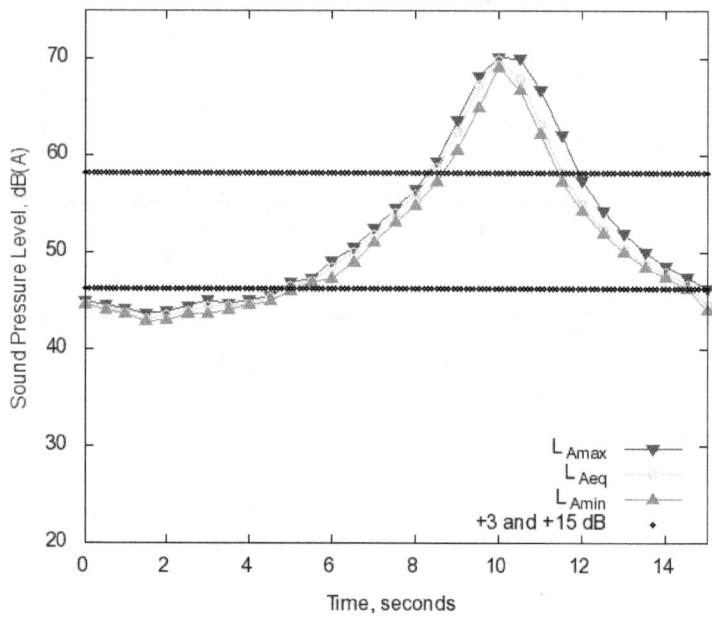

Figure A-5. Prius Time History for 30 mph Constant Speed Passby at 12 ft Microphone

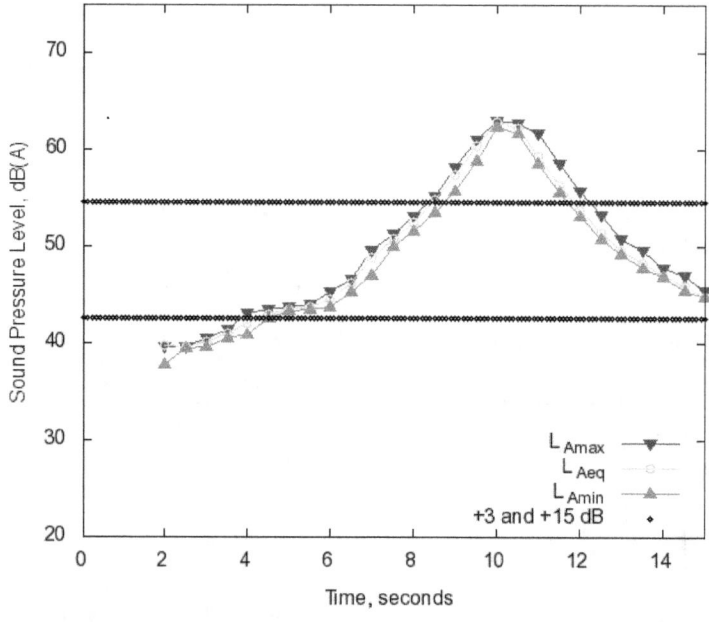

Figure A-6. Prius Time History for Acceleration Passby at 12 ft Microphone

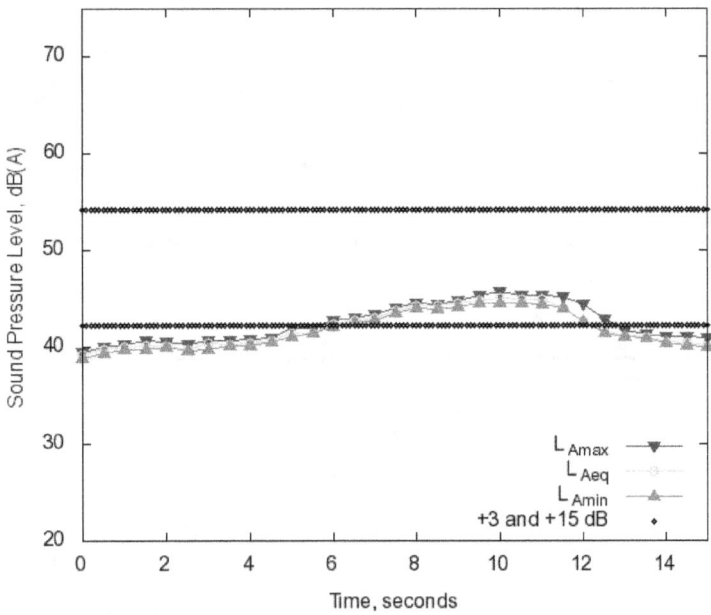

Figure A-7. Prius Time History for Reverse 5 mph Constant Speed Passby at 12 ft Microphone

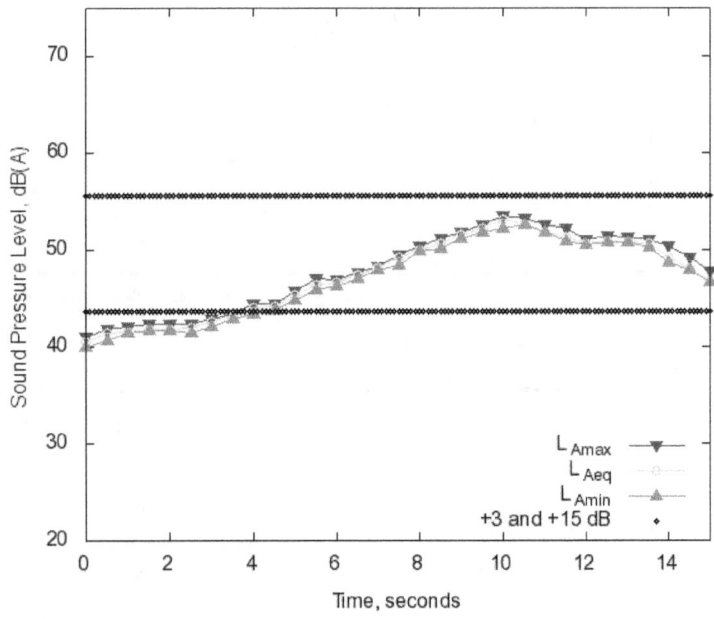

Figure A-8. Prius Time History for Deceleration Passby at 12 ft Microphone

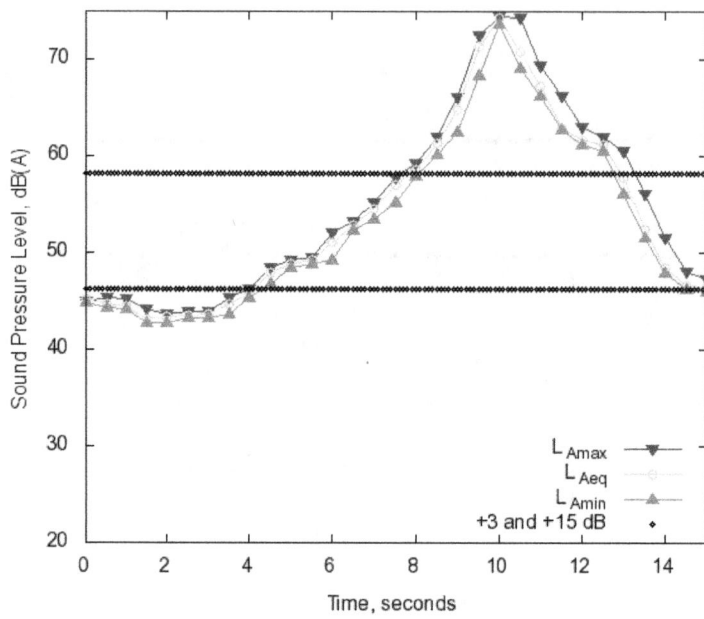

Figure A-9. Prius Time History for 40 mph Constant Speed Passby at 12 ft Microphone

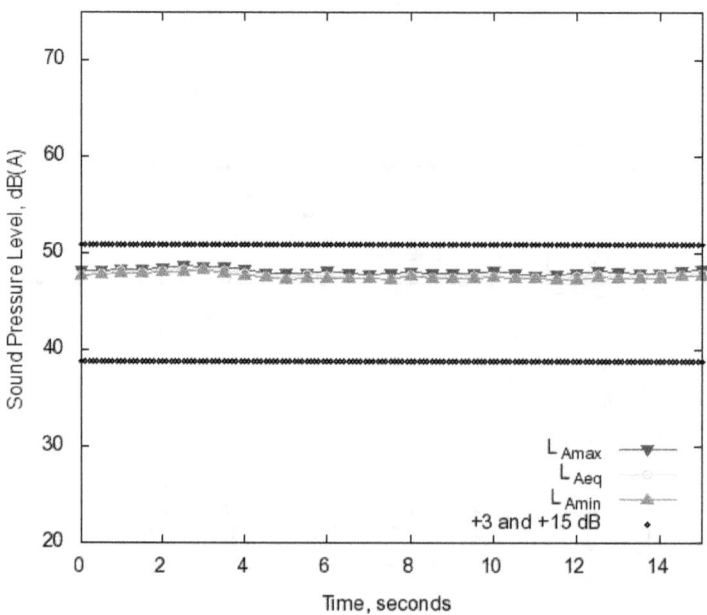

Figure A-10. Matrix Time History for Idle at 12 ft Microphone

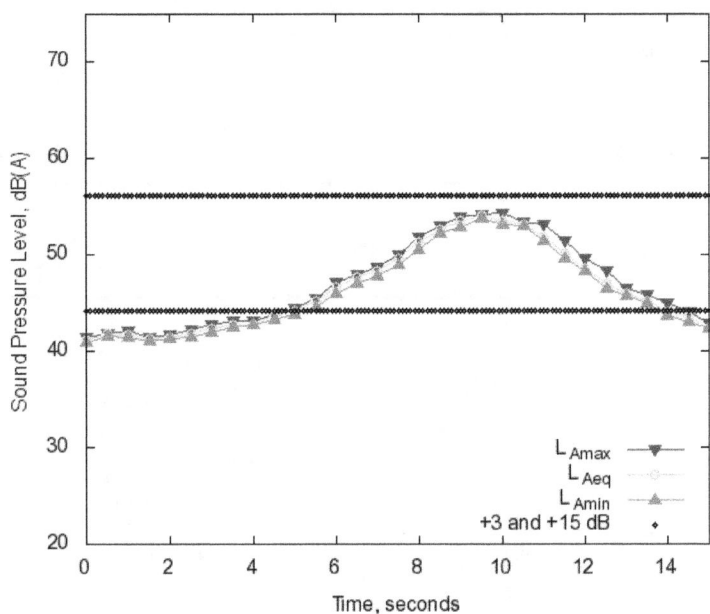

Figure A-11. Matrix Time History for 6 mph Constant Speed Passby at 12 ft Microphone

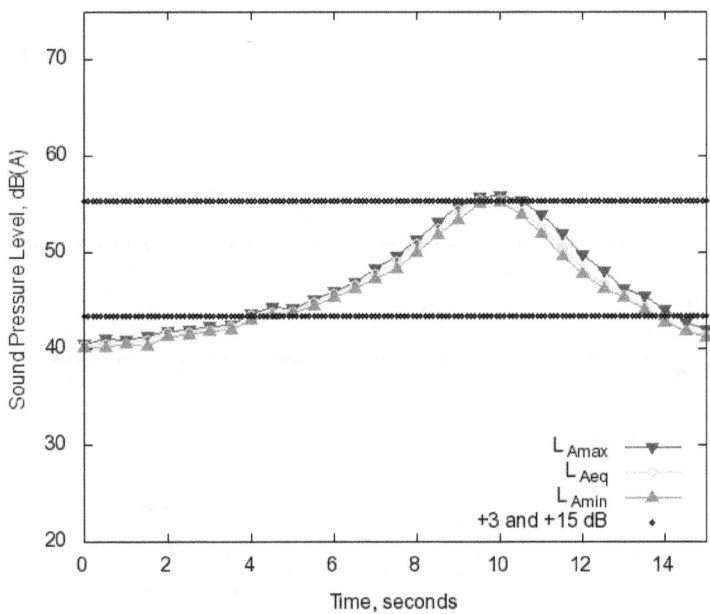

Figure A-12. Matrix Time History for 10 mph Constant Speed Passby at 12 ft Microphone

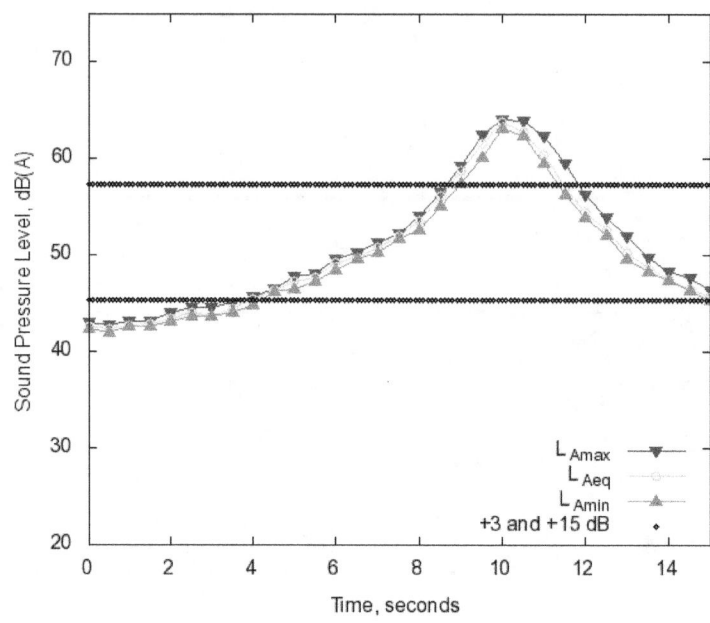

Figure A-13. Matrix Time History for 20 mph Constant Speed Passby at 12 ft Microphone

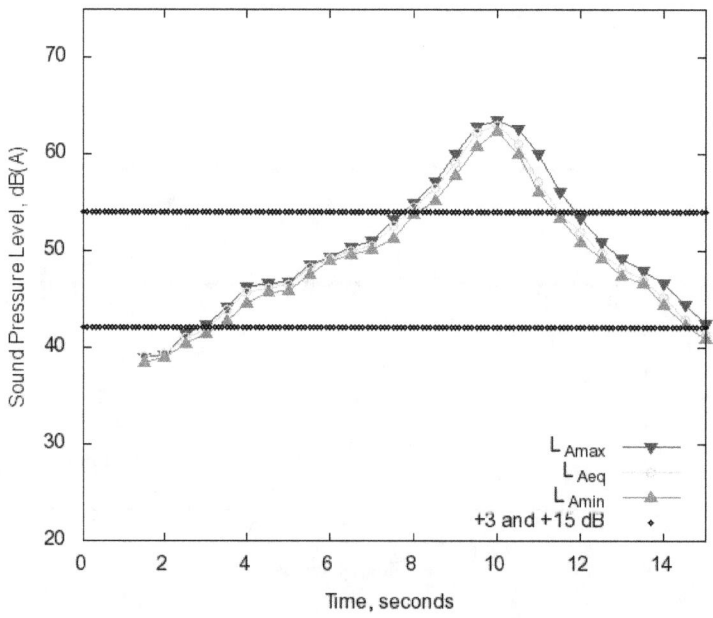

Figure A-14. Matrix Time History for 30 mph Constant Speed Passby at 12 ft Microphone

Appendix A: Acoustic Data for Vehicles

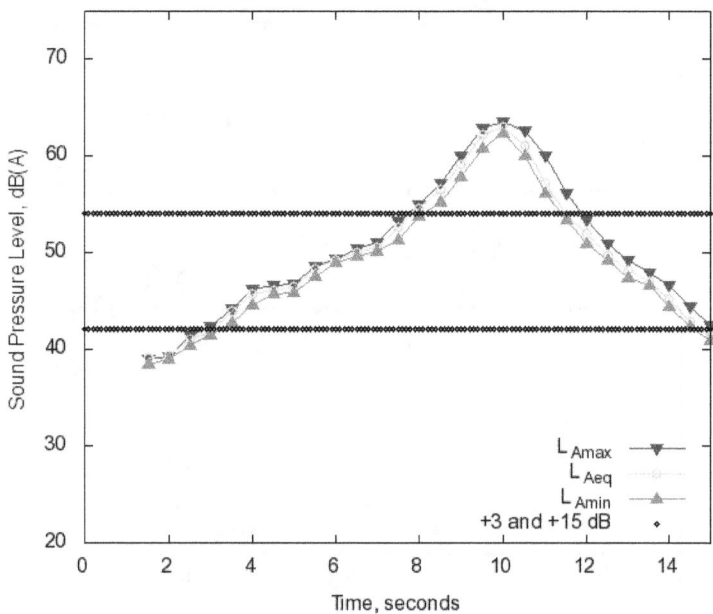

Figure A-15. Matrix Time History for Acceleration Passby at 12 ft Microphone

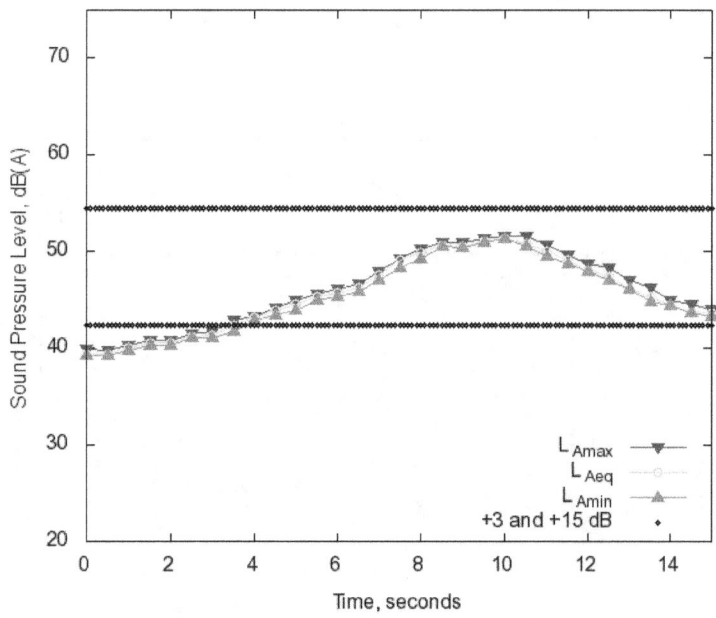

Figure A-16. Matrix Time History for Reverse 5 mph Constant Speed Passby at 12 ft Microphone

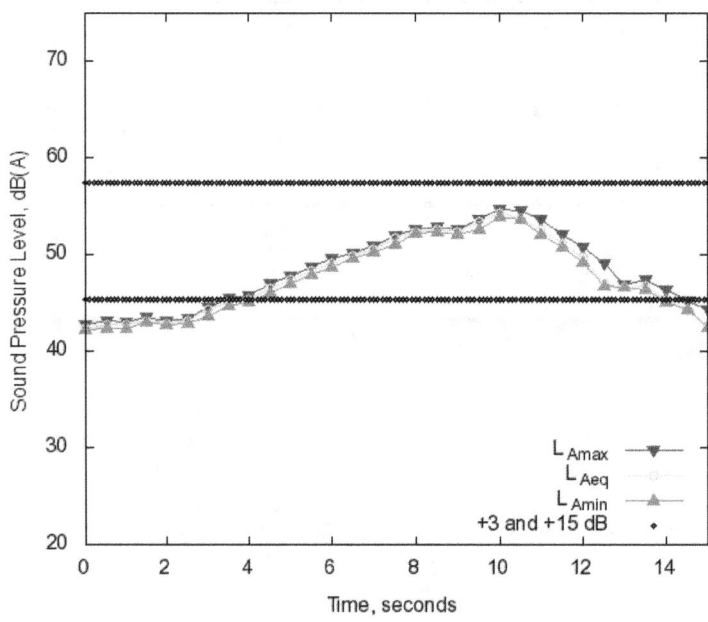

Figure A-17. Matrix Time History for Deceleration Passby at 12 ft Microphone

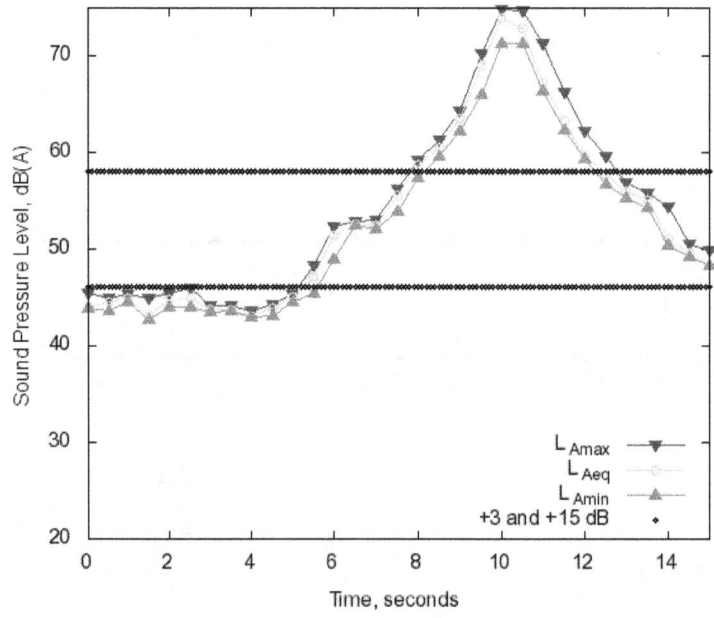

Figure A-18. Matrix Time History for 40 mph Constant Speed Passby at 12 ft Microphone

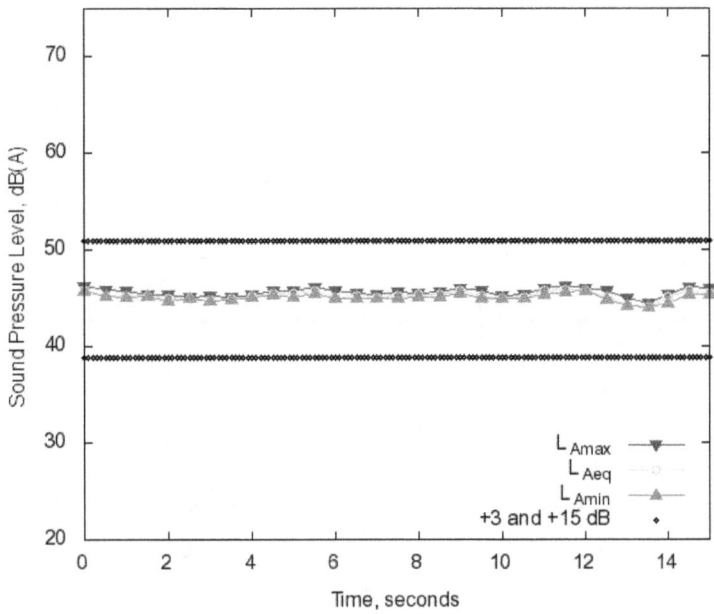

Figure A-19. Honda Civic Hybrid Time History for Idle at 12 ft Microphone

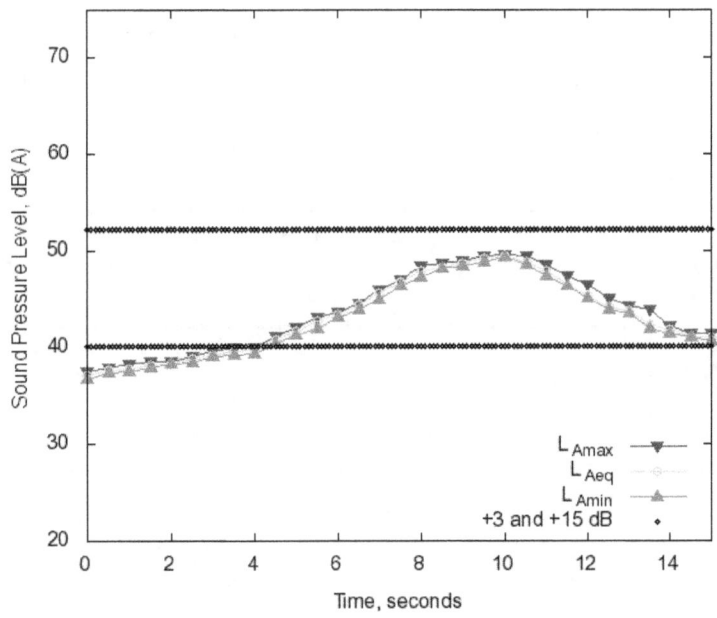

Figure A-20. Honda Civic Hybrid Time History for 6 mph Constant Speed Passby at 12 ft Microphone

Appendix A: Acoustic Data for Vehicles

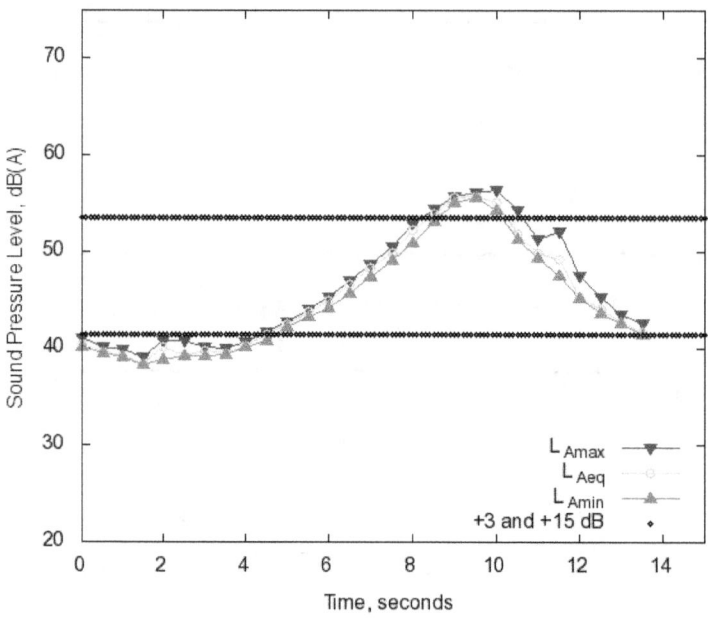

Figure A-21. Honda Civic Hybrid Time History for 10 mph Constant Speed Passby at 12 ft Microphone

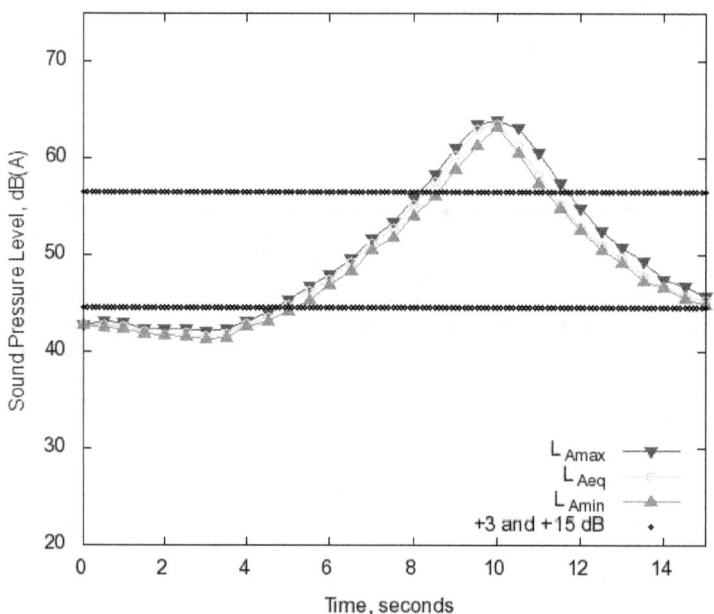

Figure A-22. Honda Civic Hybrid Time History for 20 mph Constant Speed Passby at 12 ft Microphone

Appendix A: Acoustic Data for Vehicles

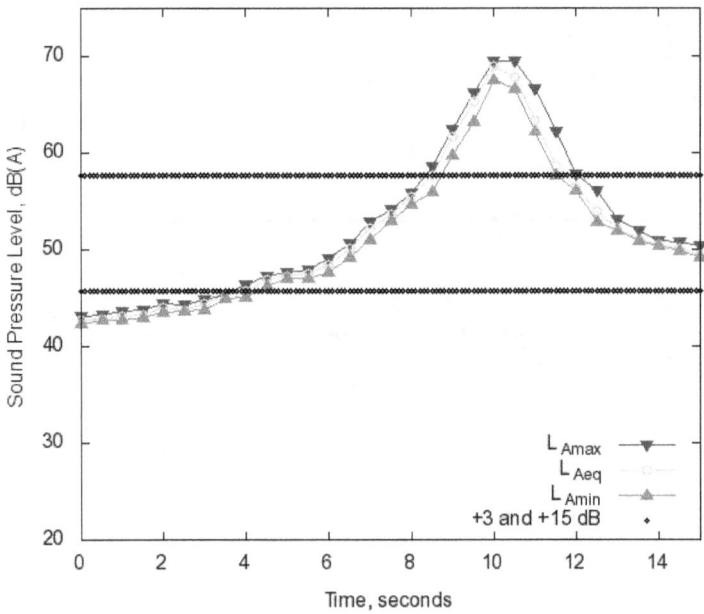

Figure A-23. Honda Civic Hybrid Time History for 30 mph Constant Speed Passby at 12 ft Microphone

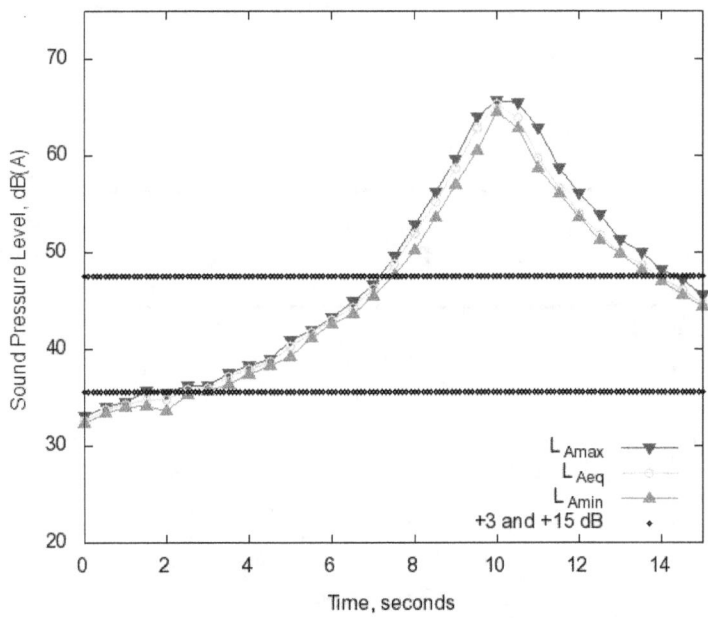

Figure A-24. Honda Civic Hybrid Time History for Acceleration Passby at 12 ft Microphone

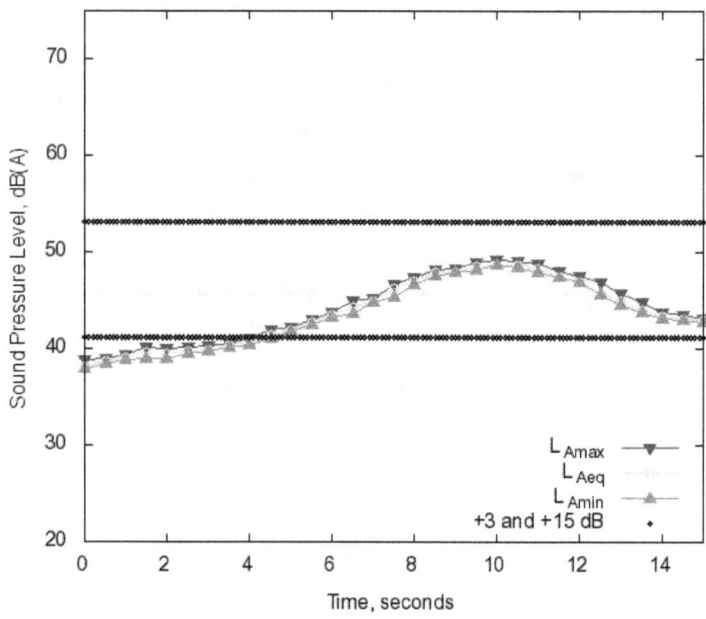

Figure A-25. Honda Civic Hybrid Time History for Reverse 5 mph Constant Speed Passby at 12 ft Microphone

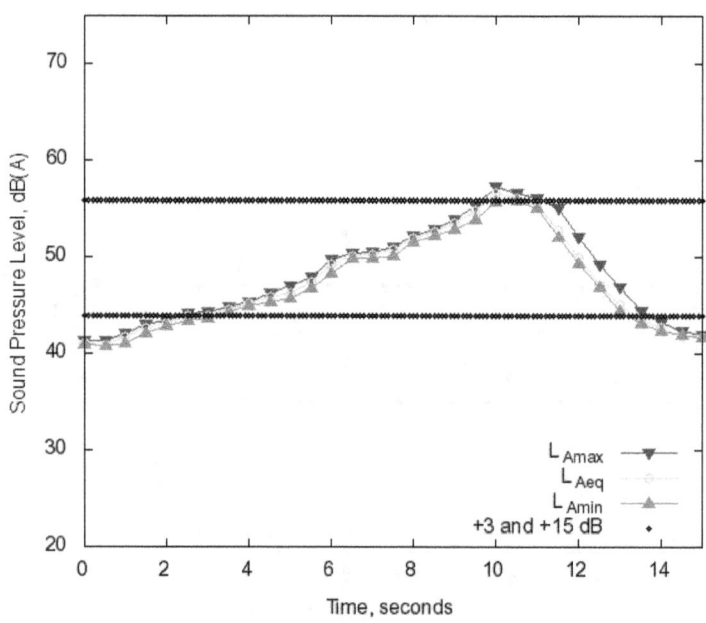

Figure A-26. Honda Civic Hybrid Time History for Deceleration Passby at 12 ft Microphone

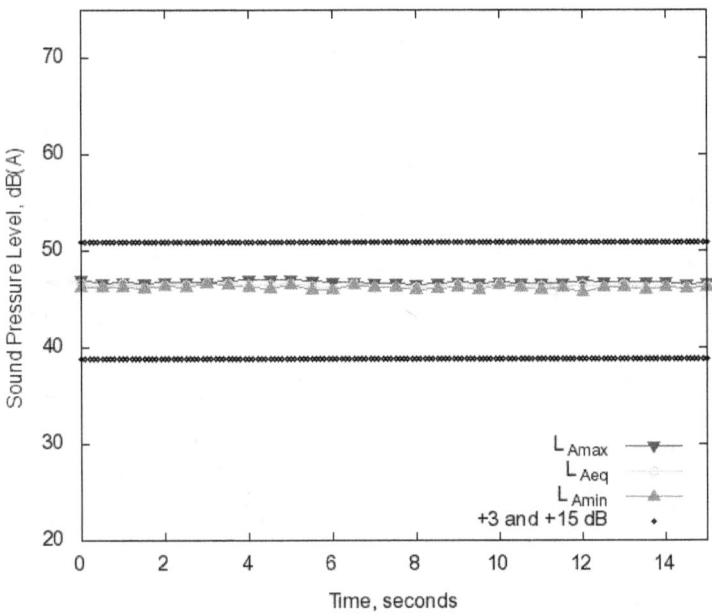

Figure A-27. Honda Civic ICE Time History for Idle at 12 ft Microphone

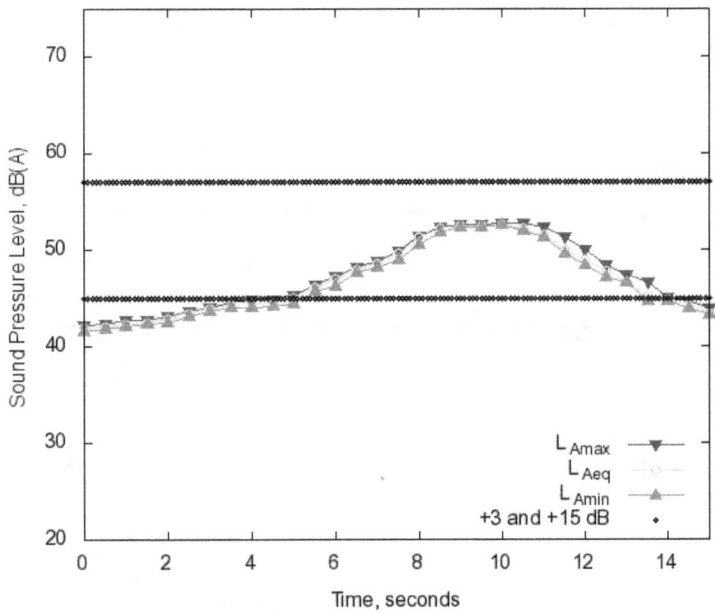

Figure A-28. Honda Civic ICE Time History for 6 mph Constant Speed Passby at 12 ft Microphone

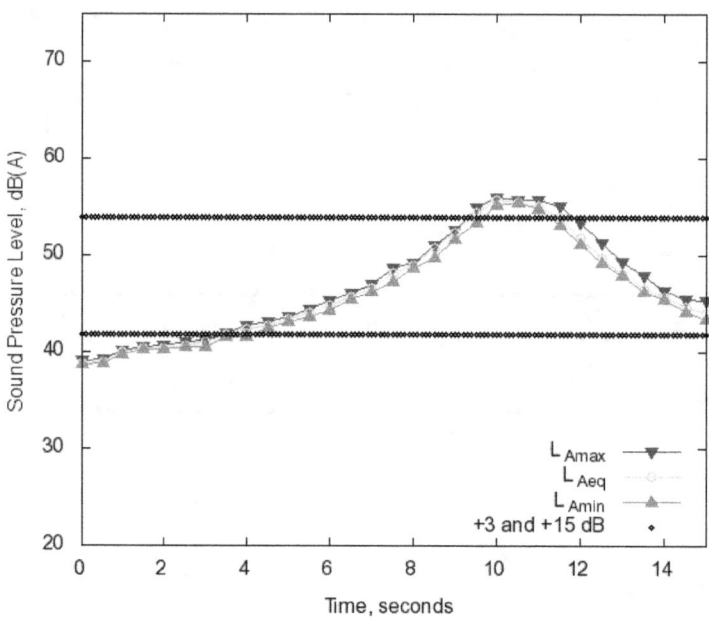

Figure A-29. Honda Civic ICE Time History for 10 mph Constant Speed Passby at 12 ft Microphone

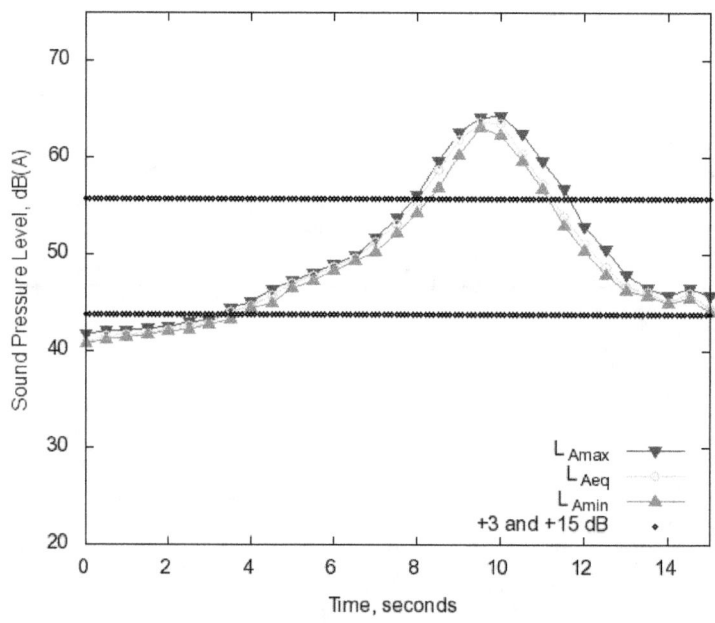

Figure A-30. Honda Civic ICE Time History for 20 mph Constant Speed Passby at 12 ft Microphone

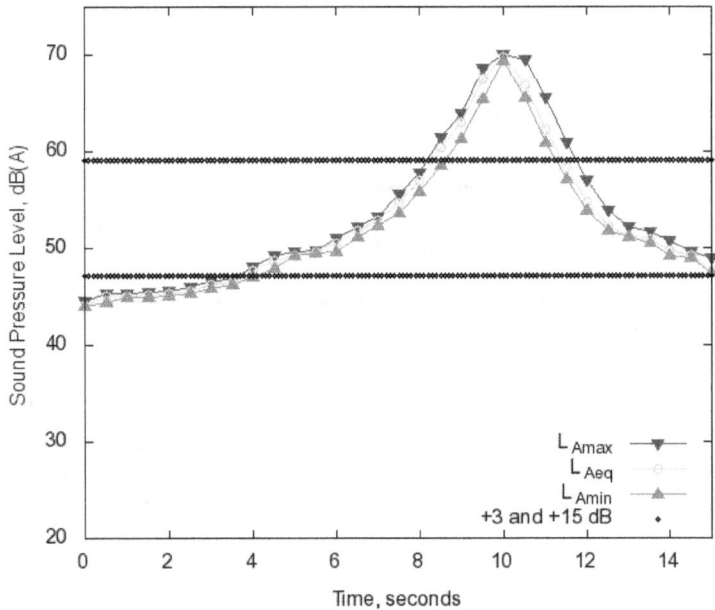

Figure A-31. Honda Civic ICE Time History for 30 mph Constant Speed Passby at 12 ft Microphone

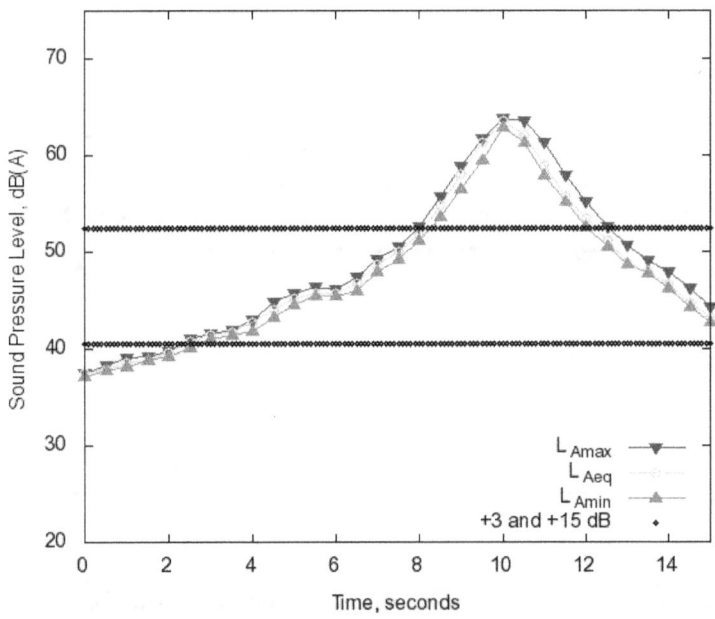

Figure A-32. Honda Civic ICE Time History for Acceleration Passby at 12 ft Microphone

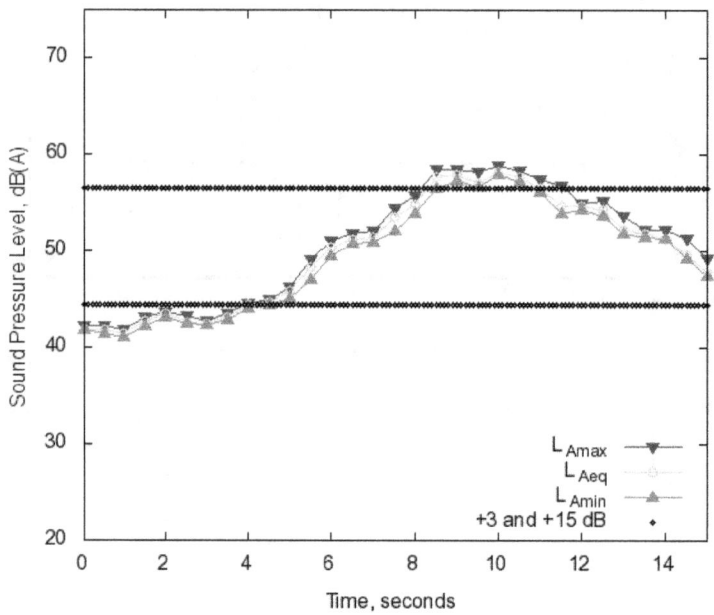

Figure A-33. Honda Civic ICE Time History for Reverse 5 mph Constant Speed Passby at 12 ft Microphone

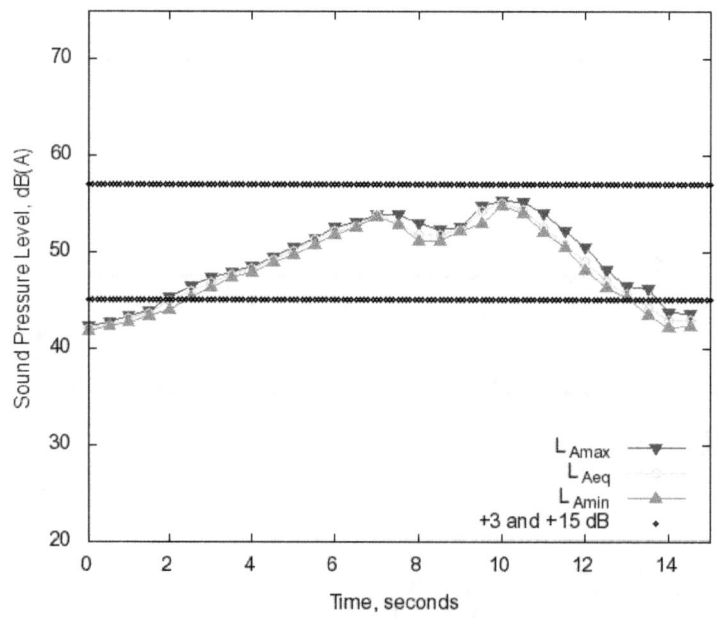

Figure A-34. Honda Civic ICE Time History for Deceleration Passby at 12 ft Microphone

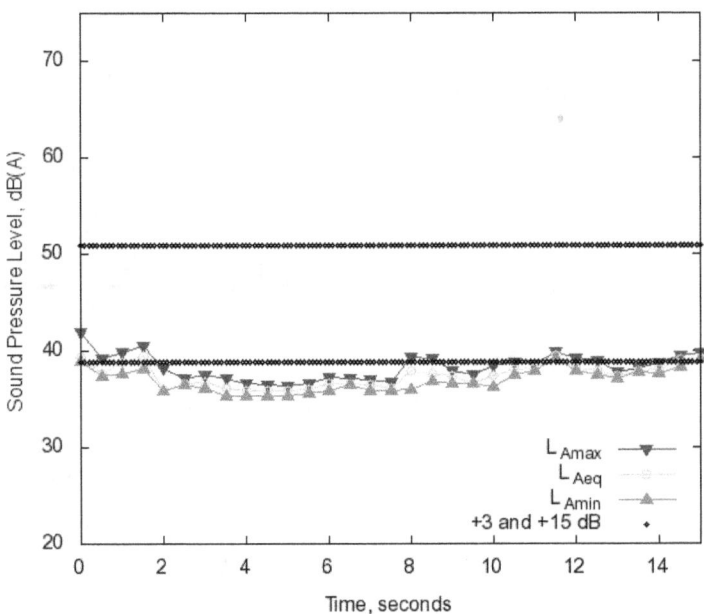

Figure A-35. Toyota Highlander Hybrid Time History for Idle at 12 ft Microphone

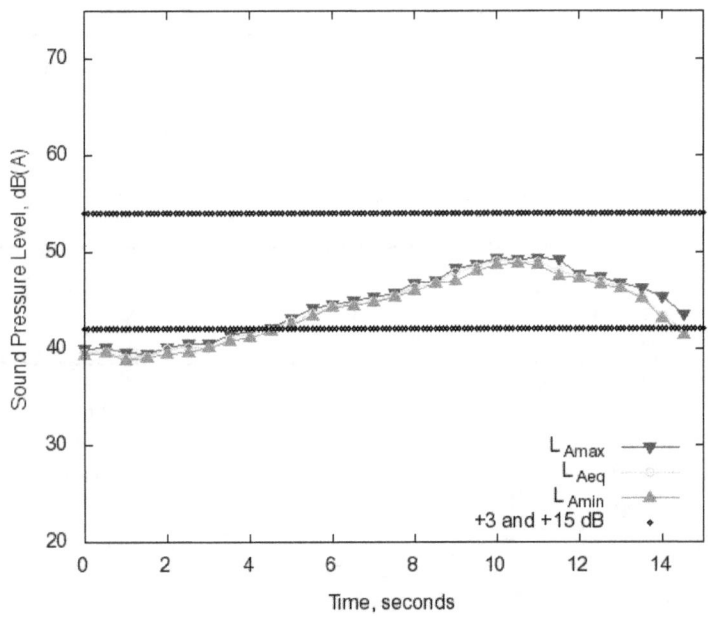

Figure A-36. Toyota Highlander Hybrid Time History for 6 mph Constant Speed Passby at 12 ft Microphone

Appendix A: Acoustic Data for Vehicles

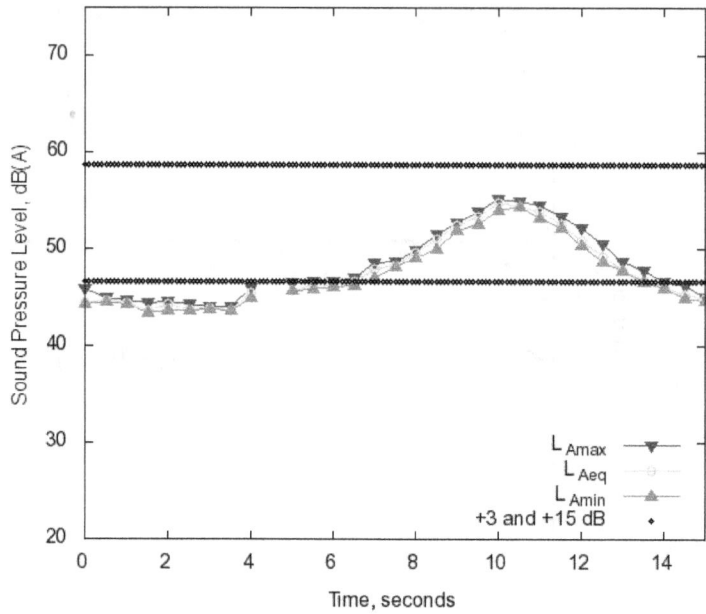

Figure A-37. Toyota Highlander Hybrid Time History for 10 mph Constant Speed Passby at 12 ft Microphone

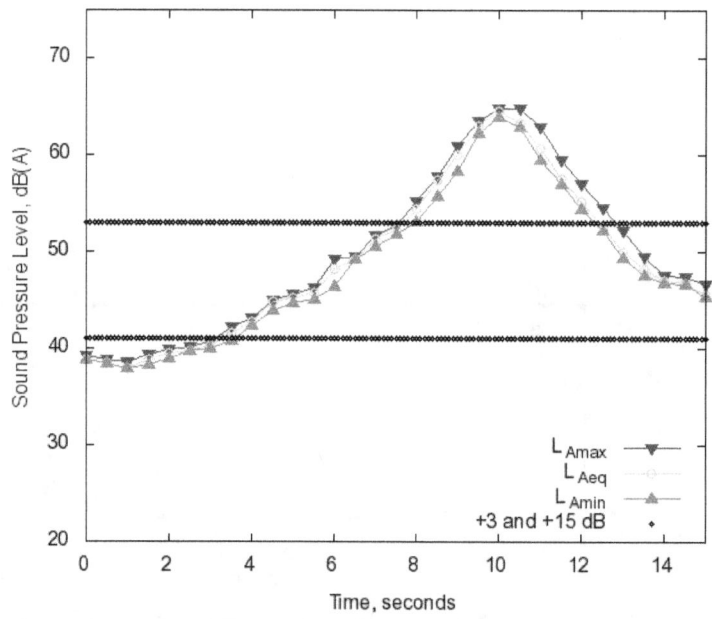

Figure A-38. Toyota Highlander Hybrid Time History for 20 mph Constant Speed Passby at 12 ft Microphone

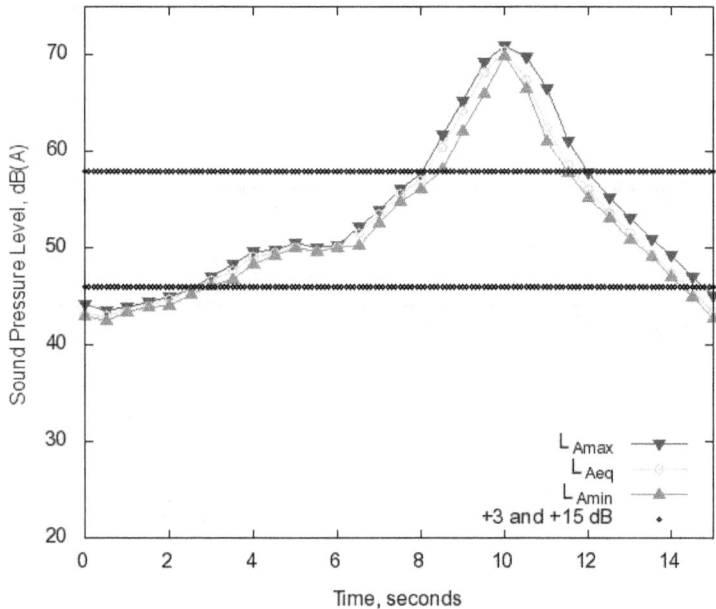

Figure A-39. Toyota Highlander Hybrid Time History for 30 mph Constant Speed Passby at 12 ft Microphone

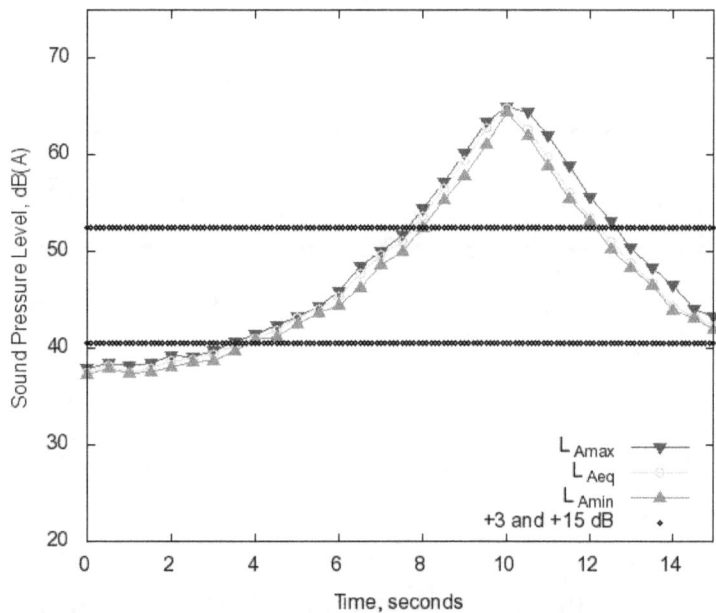

Figure A-40. Toyota Highlander Hybrid Time History for Acceleration Passby at 12 ft Microphone

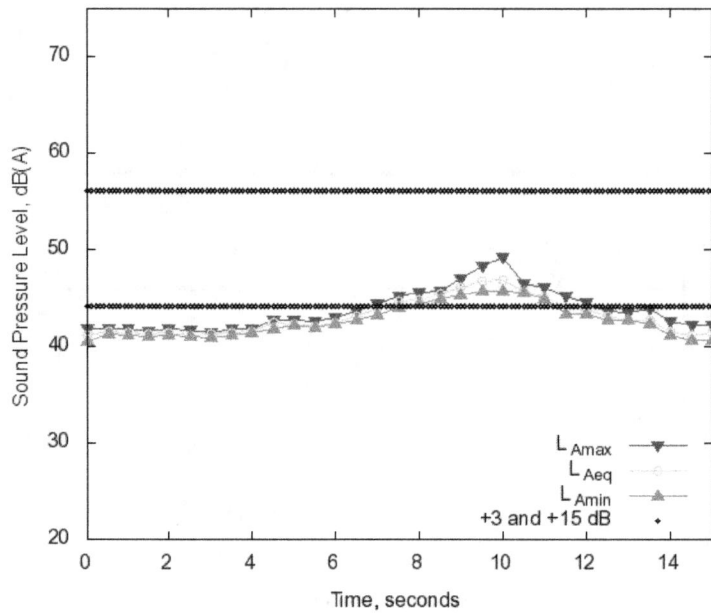

Figure A-41. Toyota Highlander Hybrid Time History for Reverse 5 mph Constant Speed Passby at 12 ft Microphone

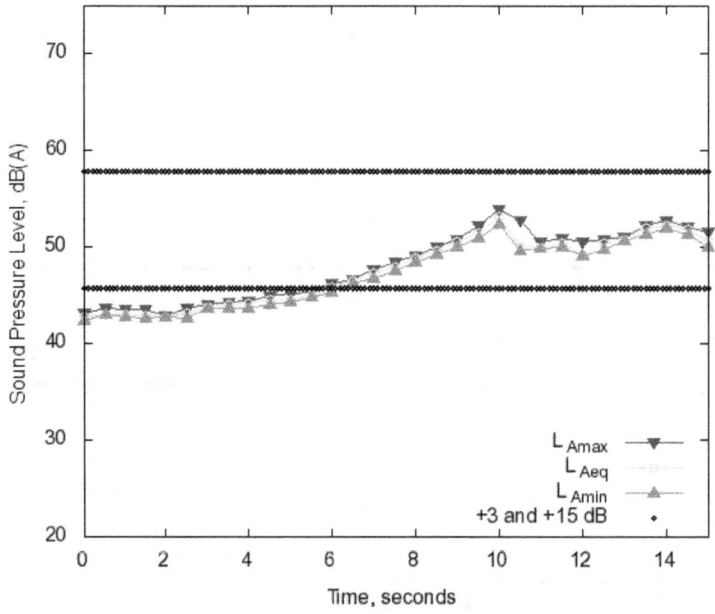

Figure A-42. Toyota Highlander Hybrid Time History for Deceleration Passby at 12 ft Microphone

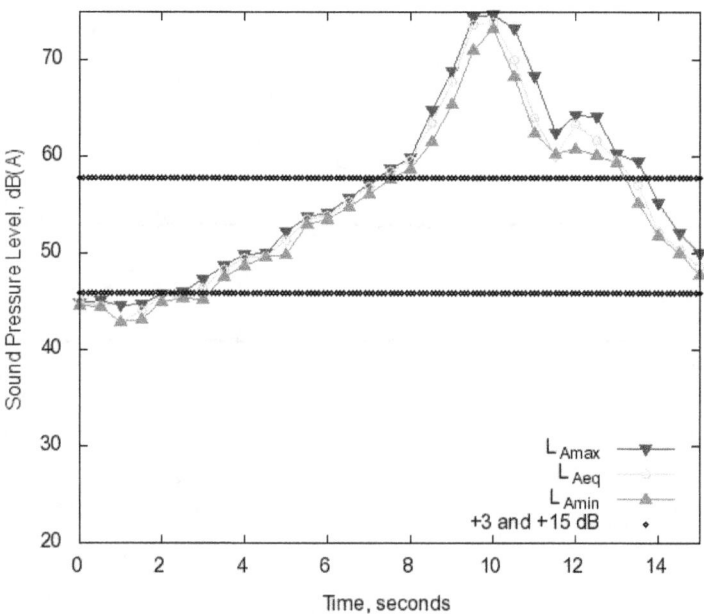

Figure A-43. Toyota Highlander Hybrid Time History for 40 mph Constant Speed Passby at 12 ft Microphone

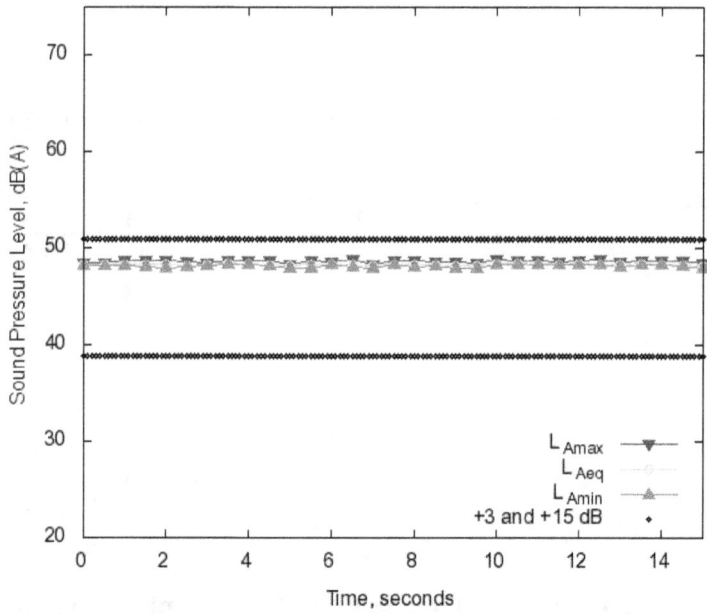

Figure A-44. Toyota Highlander ICE Time History for Idle at 12 ft Microphone

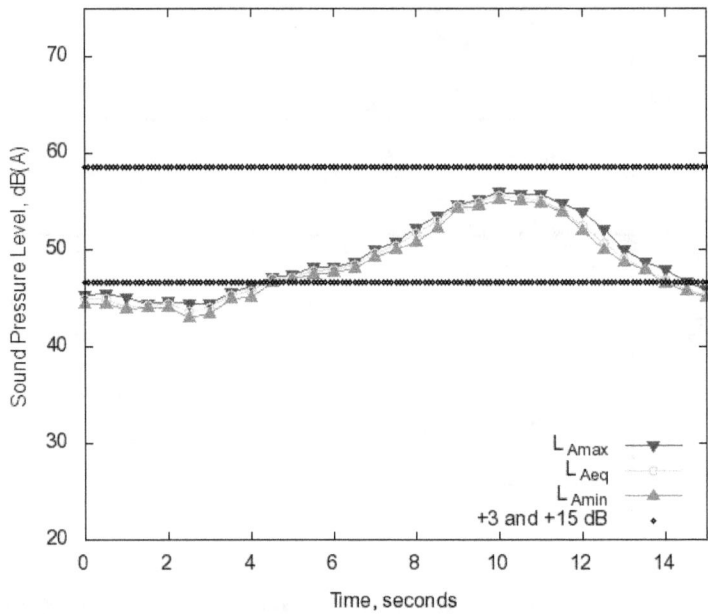

Figure A-45. Toyota Highlander ICE Time History for 6 mph Constant Speed Passby at 12 ft Microphone

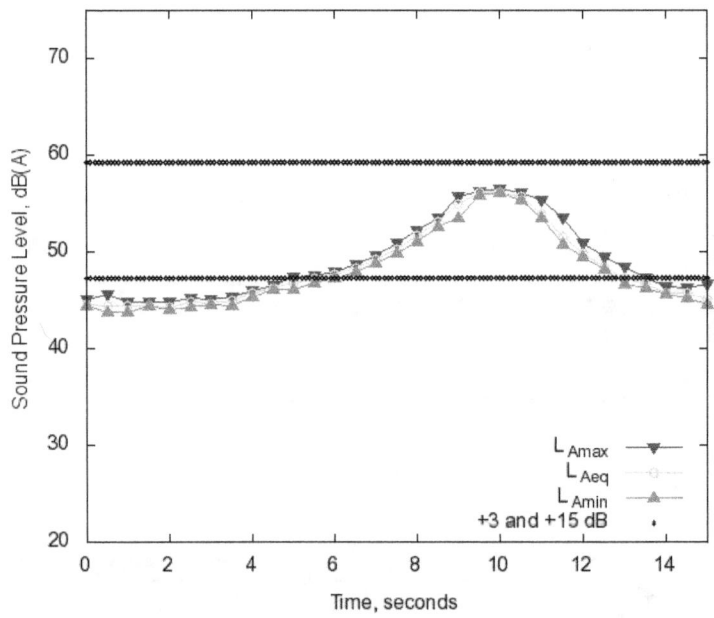

Figure A-46. Toyota Highlander ICE Time History for 10 mph Constant Speed Passby at 12 ft Microphone

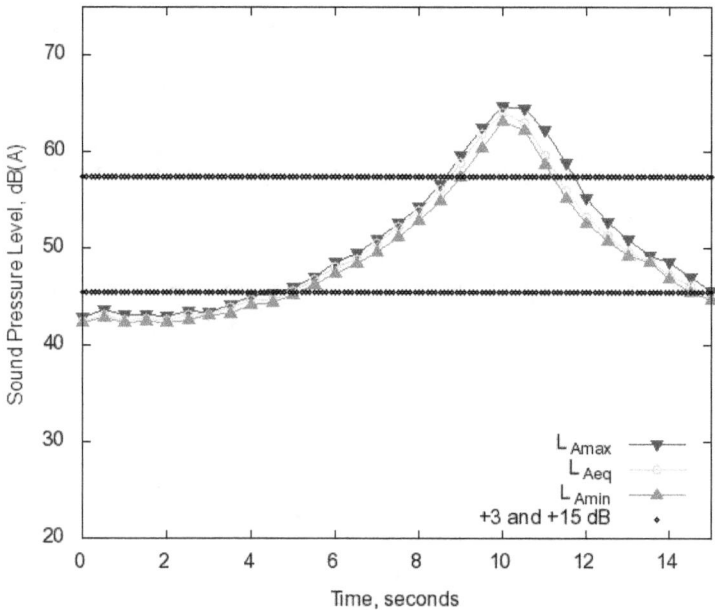

Figure A-47. Toyota Highlander ICE Time History for 20 mph Constant Speed Passby at 12 ft Microphone

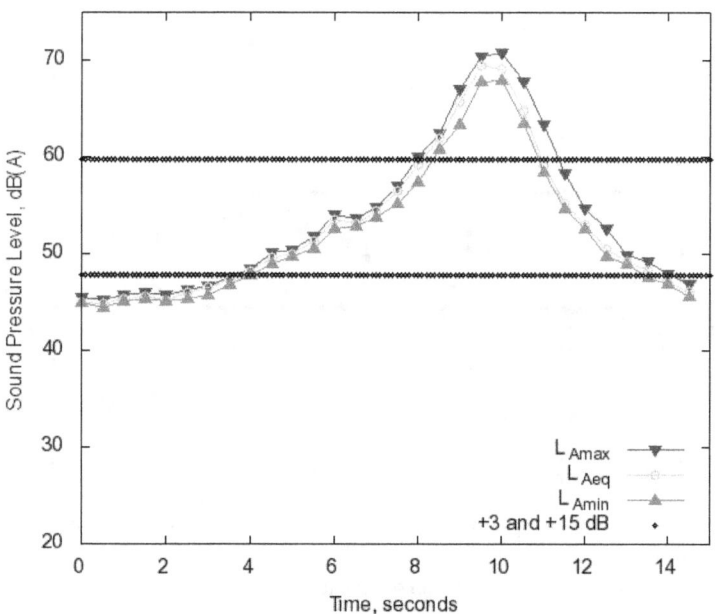

Figure A-48. Toyota Highlander ICE Time History for 30 mph Constant Speed Passby at 12 ft Microphone

Appendix A: Acoustic Data for Vehicles

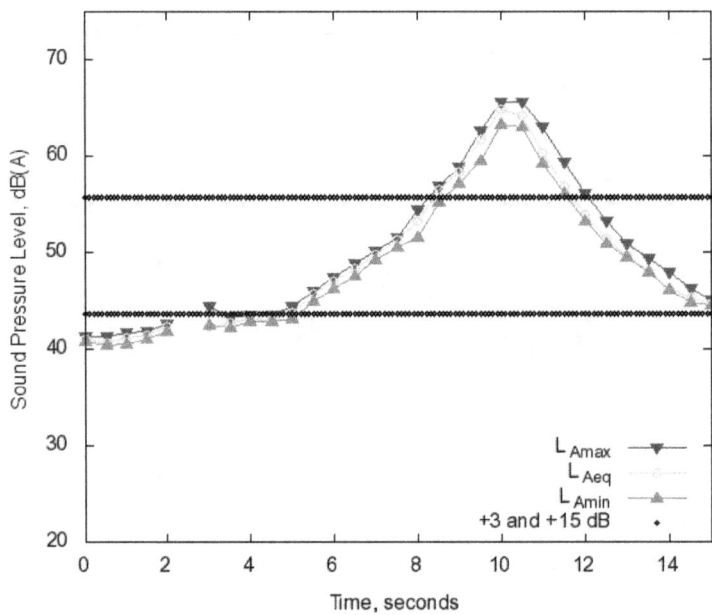

Figure A-49. Toyota Highlander ICE Time History for Acceleration Passby at 12 ft Microphone

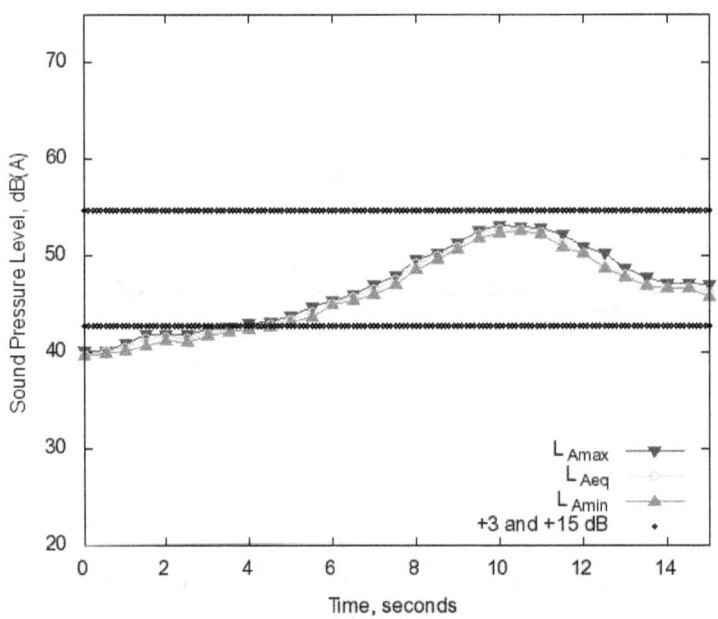

Figure A-50. Toyota Highlander ICE Time History for Reverse 5 mph Constant Speed Passby at 12 ft Microphone

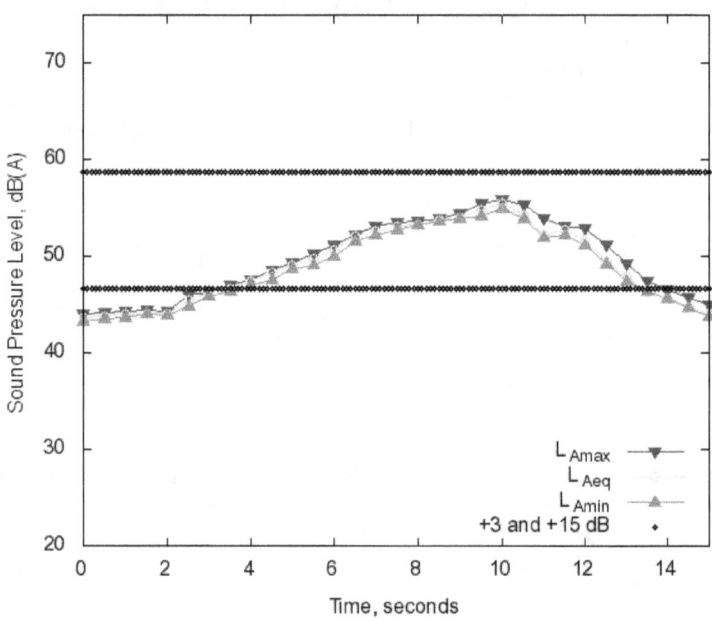

Figure A-51. Toyota Highlander ICE Time History for Deceleration Passby at 12 ft Microphone

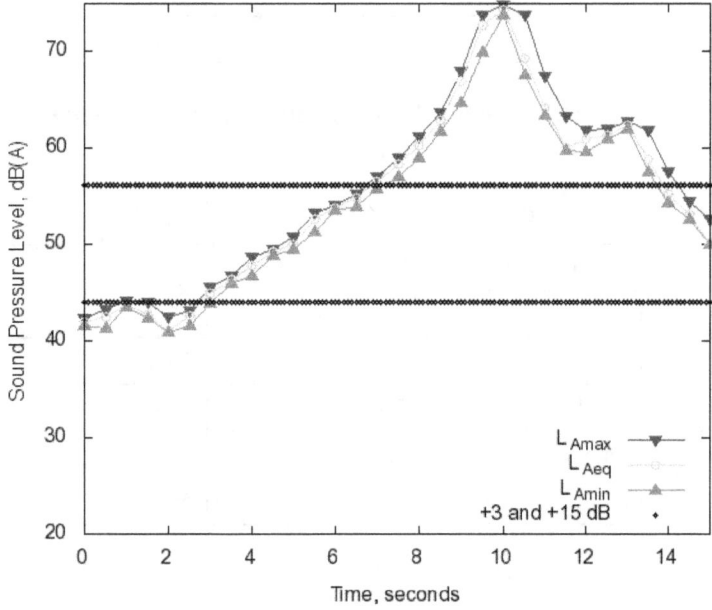

Figure A-52. Toyota Highlander ICE Time History for 40 mph Constant Speed Passby at 12 ft Microphone

A.2 Overall Levels (12-ft Microphone)

Table A-1. Overall Levels by Vehicle for Idle at 12 ft Microphone

Vehicle Type	Operation	L_{Amin}	L_{Aeq}	L_{Amax}
Prius	Idle	Background	Background	Background
Matrix	Idle	47.6	47.8	48.1
Civic Hybrid	Idle	44.6	44.8	45.1
Civic	Idle	45.8	46.0	46.4
Highlander Hybrid	Idle	Background	Background	Background
Highlander	Idle	47.9	48.1	48.5

Table A-2. Overall Levels by Vehicle for 6 mph Constant Speed Passby at 12 ft Microphone

Vehicle Type	Operation	L_{Amin}	L_{Aeq}	L_{Amax}
Prius	6 mph	44.4	44.7	45.1
Matrix	6 mph	53.0	53.5	54.2
Civic Hybrid	6 mph	49.2	49.3	49.5
Civic	6 mph	51.8	52.0	52.6
Highlander Hybrid	6 mph	52.5	53.2	54.9
Highlander	6 mph	55.2	55.5	55.9

Table A-3. Overall Levels by Vehicle for 10 mph Constant Speed Passby at 12 ft Microphone

Vehicle Type	Operation	L_{Amin}	L_{Aeq}	L_{Amax}
Prius	10 mph	52.9	53.0	53.3
Matrix	10 mph	55.2	55.4	55.8
Civic Hybrid	10 mph	54.2	55.0	56.4
Civic	10 mph	55.1	55.6	55.9
Highlander Hybrid	10 mph	53.9	54.6	55.0
Highlander	10 mph	55.9	56.0	56.3

Table A-4. Overall Levels by Vehicle for 20 mph Constant Speed Passby at 12 ft Microphone

Vehicle Type	Operation	L_{Amin}	L_{Aeq}	L_{Amax}
Prius	20 mph	62.8	63.0	63.3
Matrix	20 mph	63.2	63.8	64.1
Civic Hybrid	20 mph	63.3	63.6	64.0
Civic	20 mph	62.4	63.5	64.4
Highlander Hybrid	20 mph	63.9	64.6	64.9
Highlander	20 mph	63.2	64.1	64.7

Table A-5. Overall Levels by Vehicle for 30 mph Constant Speed Passby at 12 ft Microphone

Vehicle Type	Operation	L_{Amin}	L_{Aeq}	L_{Amax}
Prius	30 mph	69.2	69.9	70.2
Matrix	30 mph	68.3	69.3	70.4
Civic Hybrid	30 mph	66.7	67.8	69.5
Civic	30 mph	69.4	69.8	70.1
Highlander Hybrid	30 mph	69.9	70.4	71.0
Highlander	30 mph	68.0	69.1	70.8

Table A-6. Overall Levels by Vehicle for 40 mph Constant Speed Passby at 12 ft Microphone

Vehicle Type	Operation	L_{Amin}	L_{Aeq}	L_{Amax}
Prius	40 mph	73.7	74.4	74.5
Matrix	40 mph	71.4	73.9	74.9
Civic Hybrid	40 mph	N/A	N/A	N/A
Civic	40 mph	N/A	N/A	N/A
Highlander Hybrid	40 mph	73.3	73.9	74.8
Highlander	40 mph	73.8	74.4	74.9

Table A-7. Overall Levels by Vehicle for Acceleration at 12 ft Microphone

Vehicle Type	Operation	L_{Amin}	L_{Aeq}	L_{Amax}
Prius	Acceleration	62.4	62.9	63.1
Matrix	Acceleration	62.4	63.1	63.6
Civic Hybrid	Acceleration	64.6	65.4	65.8
Civic	Acceleration	62.9	63.5	63.8
Highlander Hybrid	Acceleration	64.5	64.8	65.0
Highlander	Acceleration	63.3	64.9	65.6

Table A-8. Overall Levels by Vehicle for Deceleration at 12 ft Microphone

Vehicle Type	Operation	L_{Amin}	L_{Aeq}	L_{Amax}
Prius	Deceleration	52.2	53.0	53.4
Matrix	Deceleration	53.8	54.2	54.6
Civic Hybrid	Deceleration	55.7	56.6	57.2
Civic	Deceleration	54.8	55.0	55.3
Highlander Hybrid	Deceleration	52.2	53.0	53.7
Highlander	Deceleration	54.9	55.4	55.8

Table A-9. Overall Levels by Vehicle for Reverse (5 mph Constant Speed Passby) at 12 ft Microphone

Vehicle Type	Operation	L_{Amin}	L_{Aeq}	L_{Amax}
Prius	Reverse	43.7	44.2	44.8
Matrix	Reverse	51.2	51.3	51.5
Civic Hybrid	Reverse	48.5	48.5	49.0
Civic	Reverse	58.0	58.2	58.9
Highlander Hybrid	Reverse	44.6	45.9	48.6
Highlander	Reverse	52.3	52.7	53.1

A.3 One-Third Octave Levels (12-ft Microphone)

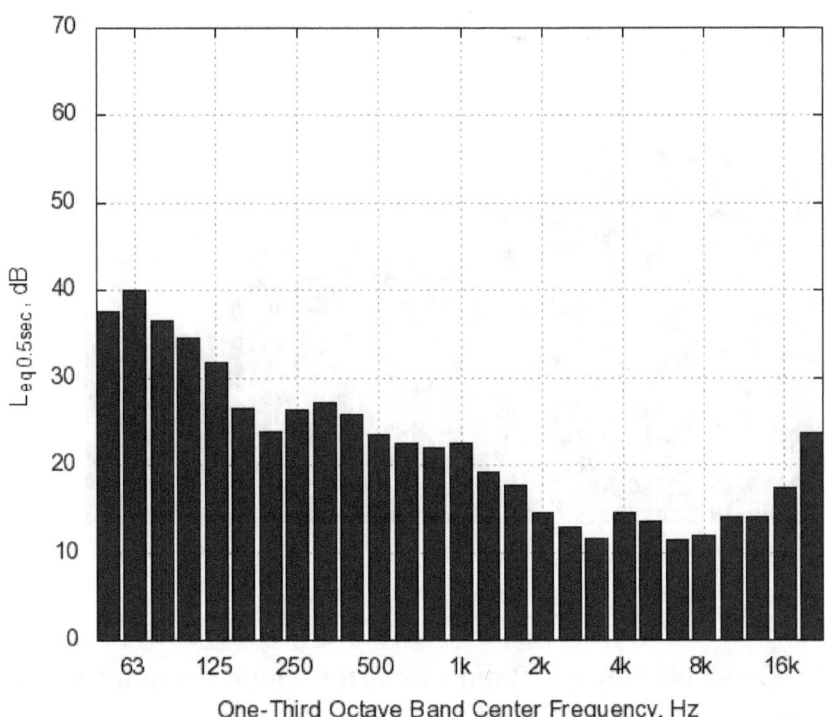

Figure A-53. Prius One-Third Octave Band Levels for Idle at 12 ft Microphone

Appendix A: Acoustic Data for Vehicles

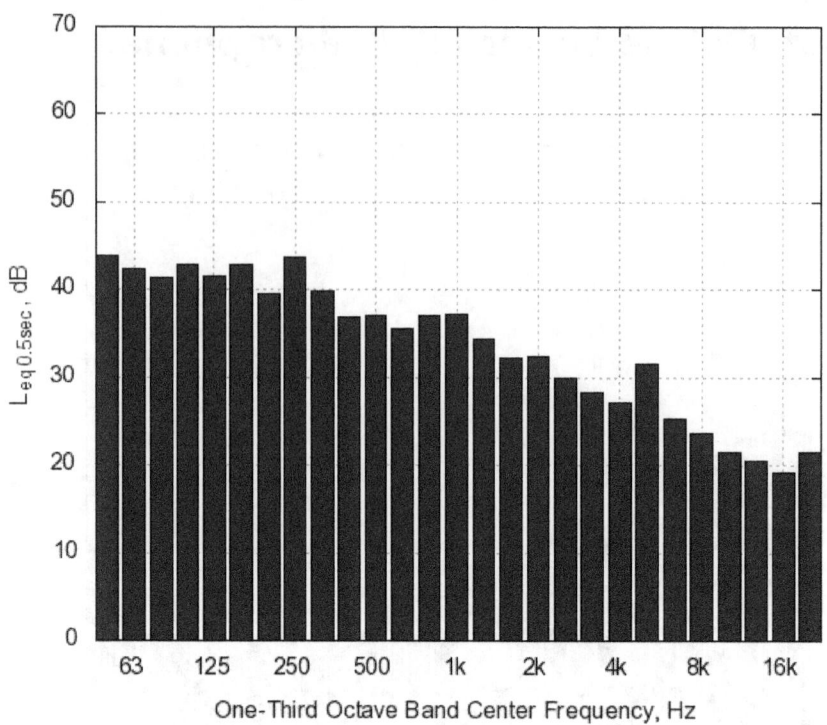

Figure A-54. Prius One-Third Octave Band Levels for 6 mph Constant Speed Passby at 12 ft Microphone

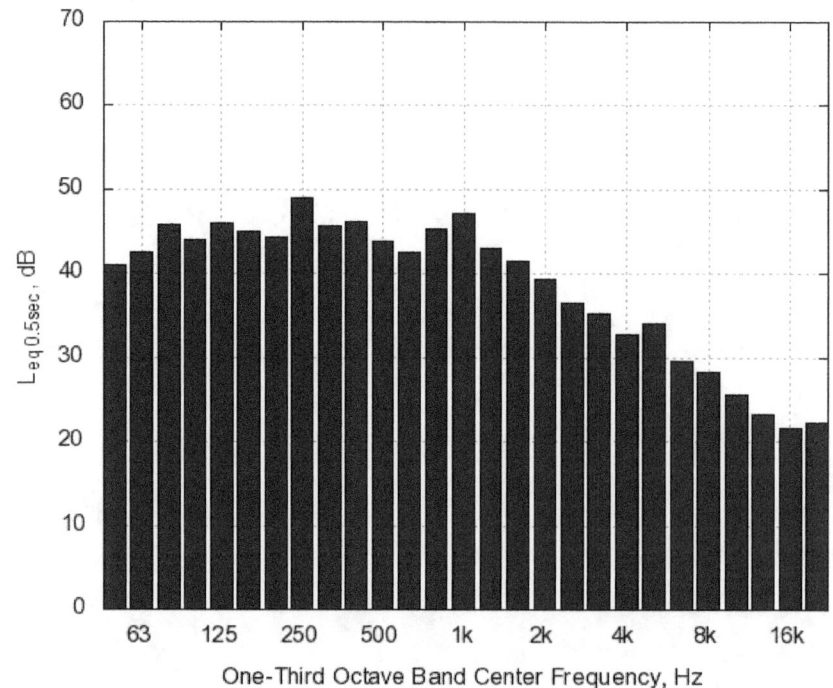

Figure A-55. Prius One-Third Octave Band Levels for 10 mph Constant Speed Passby at 12 ft Microphone

Appendix A: Acoustic Data for Vehicles

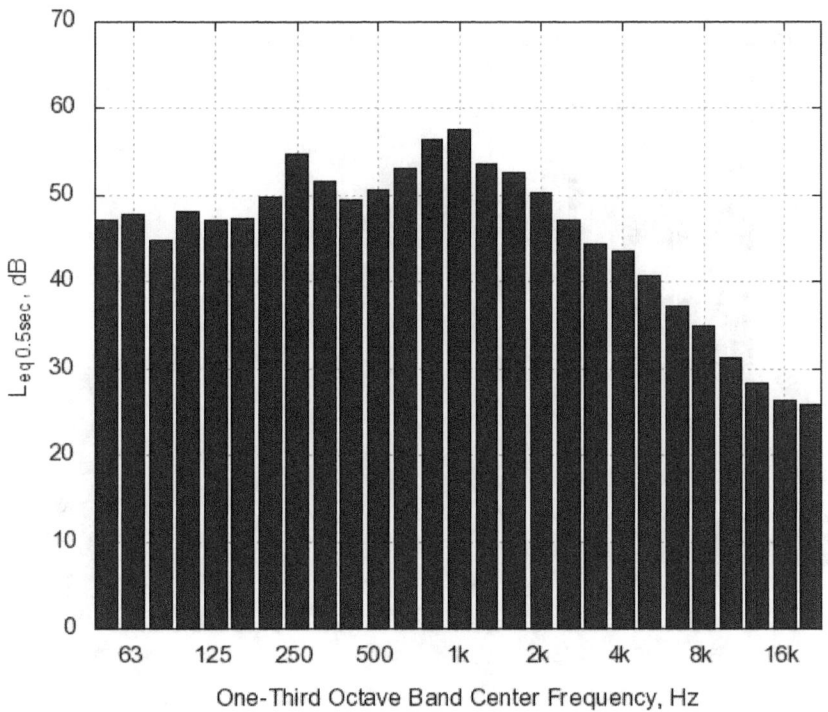

Figure A-56. Prius One-Third Octave Band Levels for 20 mph Constant Speed Passby at 12 ft Microphone

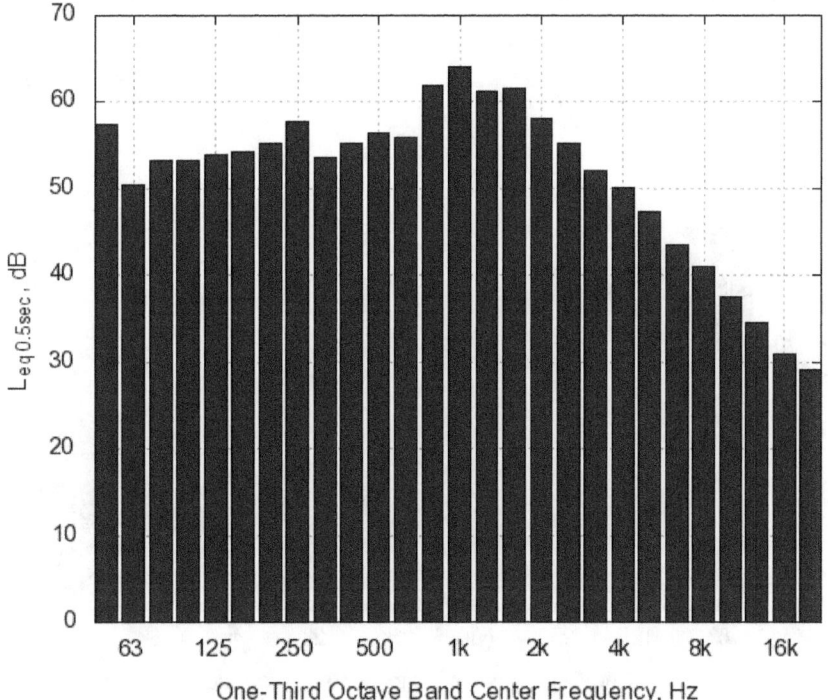

Figure A-57. Prius One-Third Octave Band Levels for 30 mph Constant Speed Passby at 12 ft Microphone

Appendix A: Acoustic Data for Vehicles

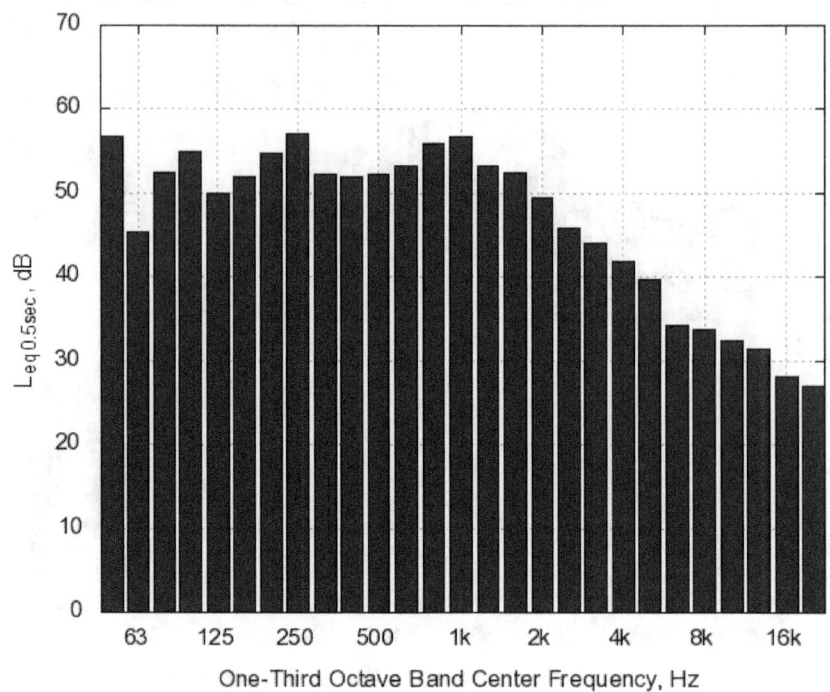

Figure A-58. Prius One-Third Octave Band Levels for Acceleration Passby at 12 ft Microphone

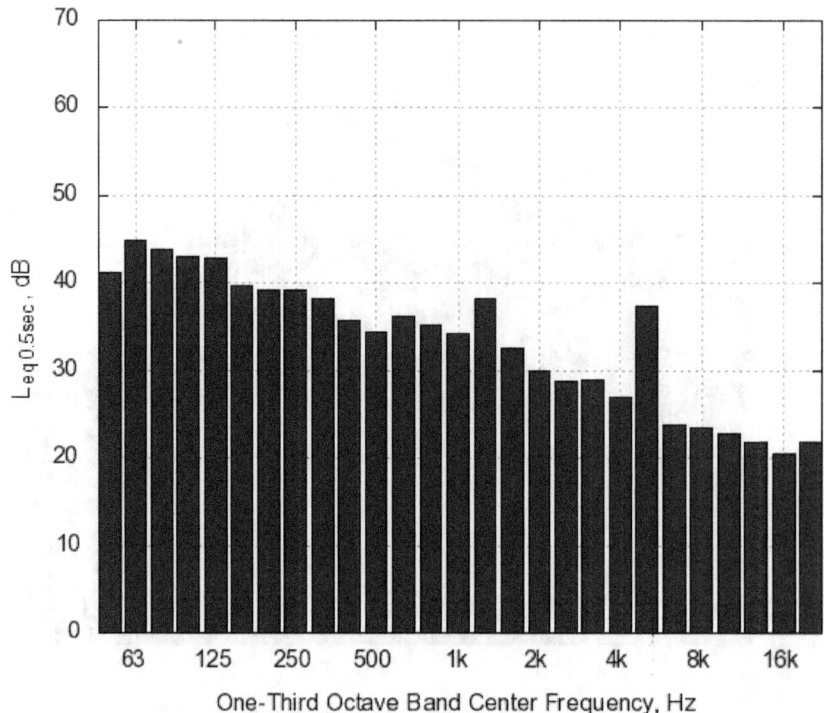

Figure A-59. Prius One-Third Octave Band Levels for Reverse 5 mph Constant Speed Passby at 12 ft Microphone

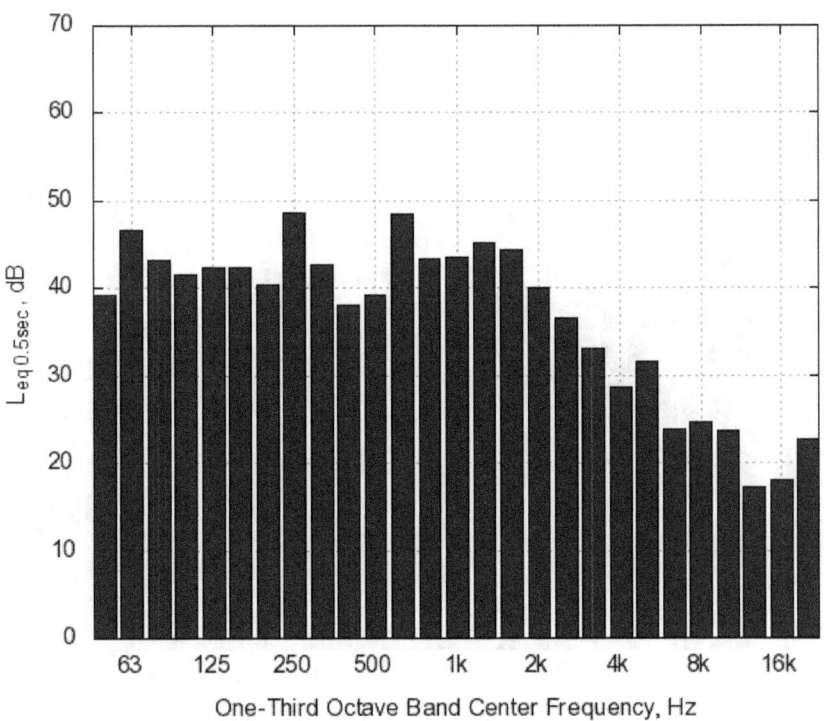

Figure A-60. Prius One-Third Octave Band Levels for Deceleration Passby at 12 ft Microphone

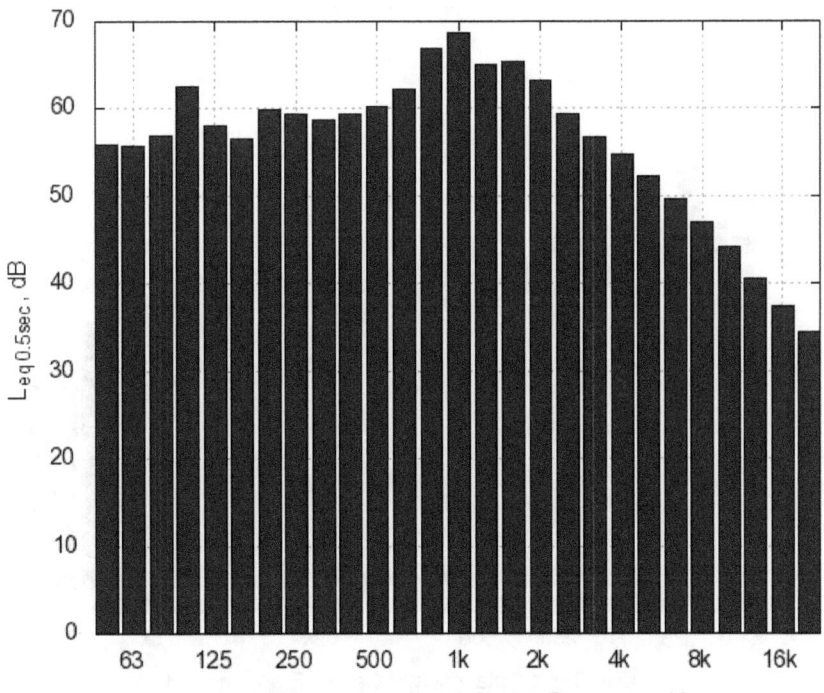

Figure A-61. Prius One-Third Octave Band Levels for 40 mph Constant Speed Passby at 12 ft Microphone

Appendix A: Acoustic Data for Vehicles

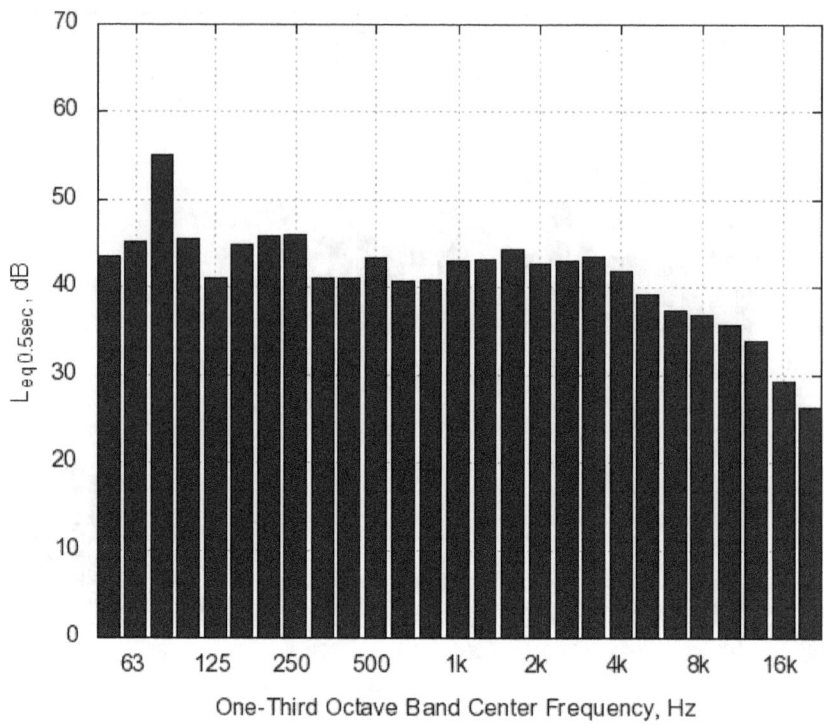

Figure A-62. Matrix One-Third Octave Band Levels for Idle at 12 ft Microphone

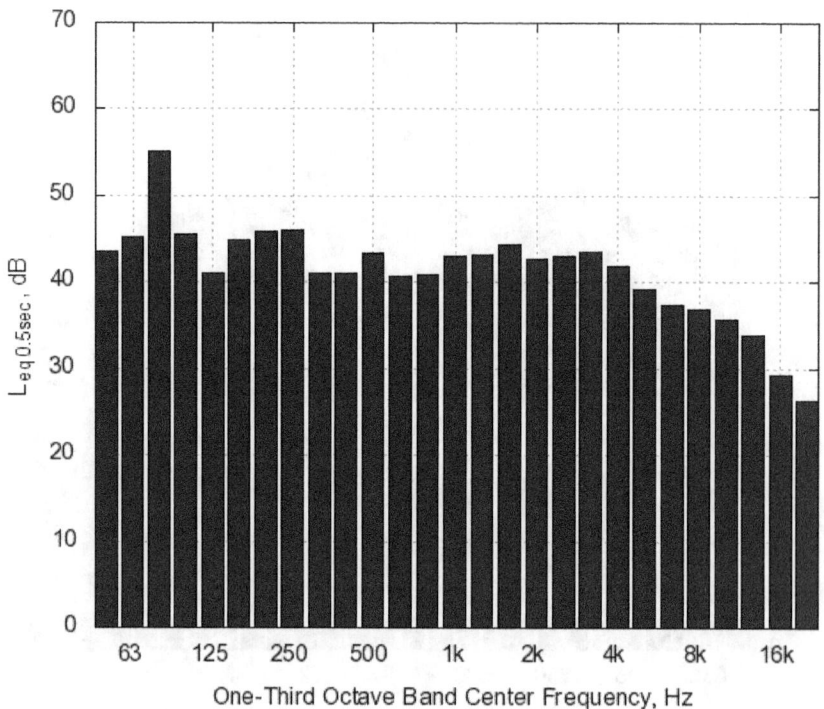

Figure A-63. Matrix One-Third Octave Band Levels for 6 mph Constant Speed Passby at 12 ft Microphone

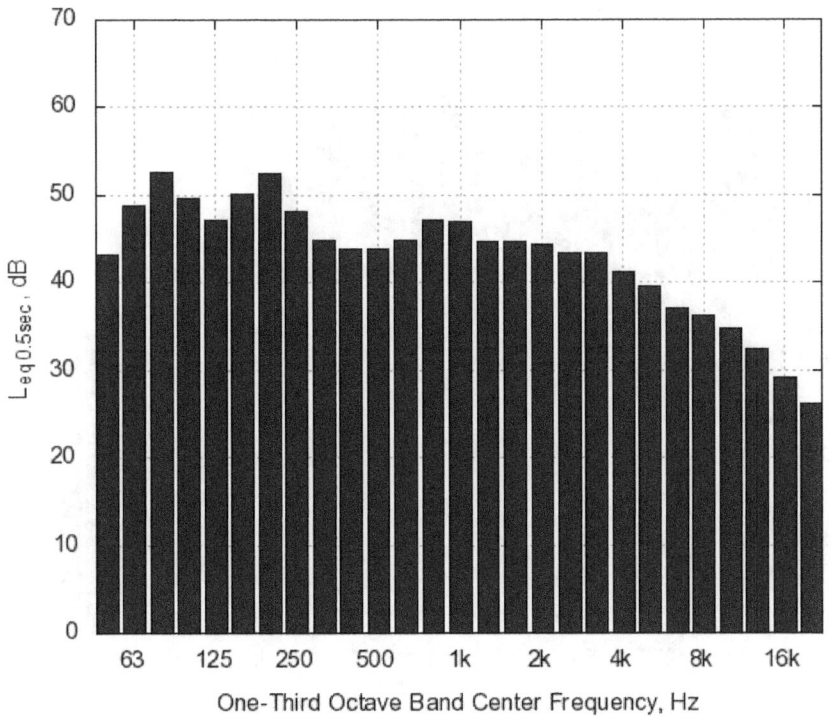

Figure A-64. Matrix One-Third Octave Band Levels for 10 mph Constant Speed Passby at 12 ft Microphone

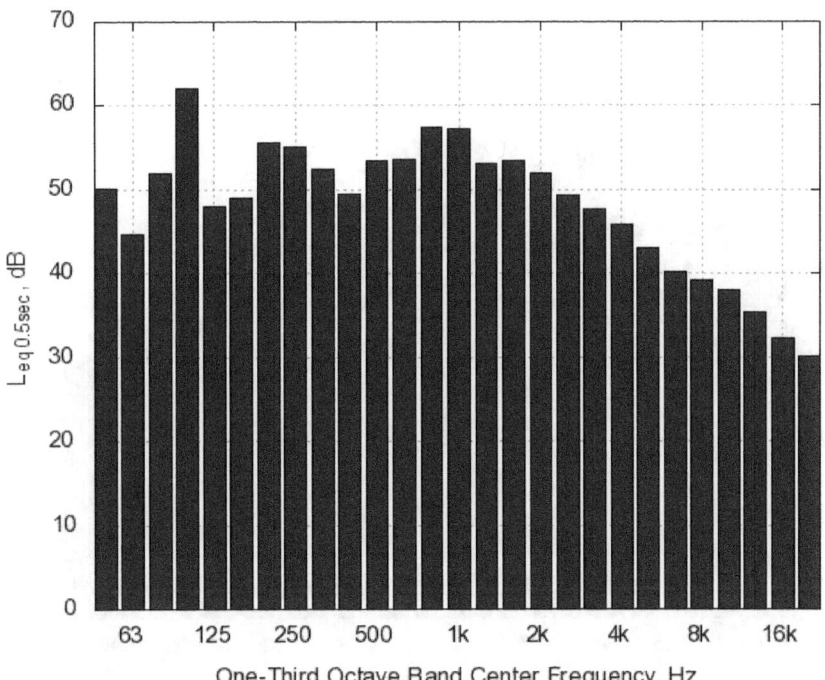

Figure A-65. Matrix One-Third Octave Band Levels for 20 mph Constant Speed Passby at 12 ft Microphone

Appendix A: Acoustic Data for Vehicles

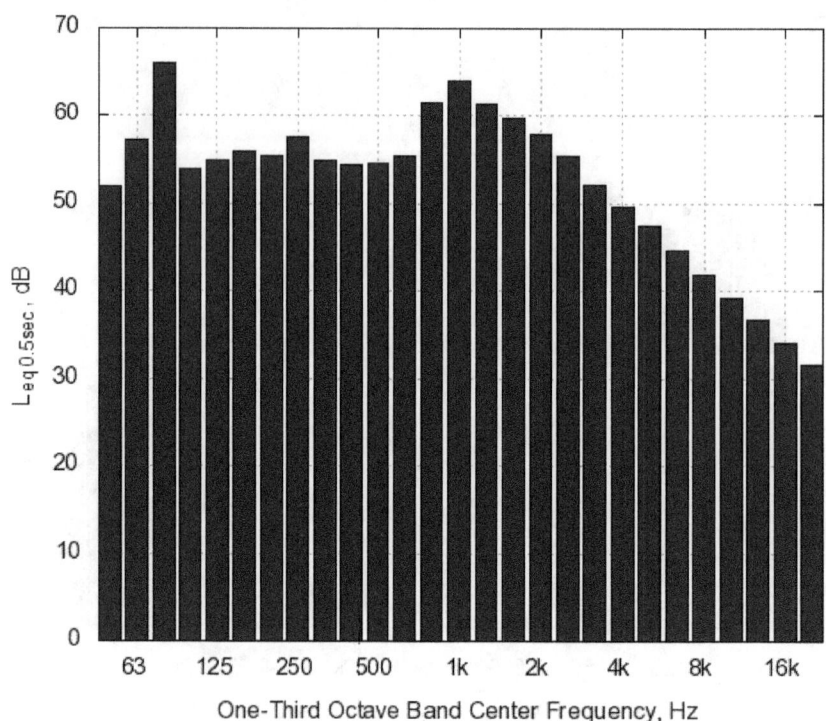

Figure A-66. Matrix One-Third Octave Band Levels for 30 mph Constant Speed Passby at 12 ft Microphone

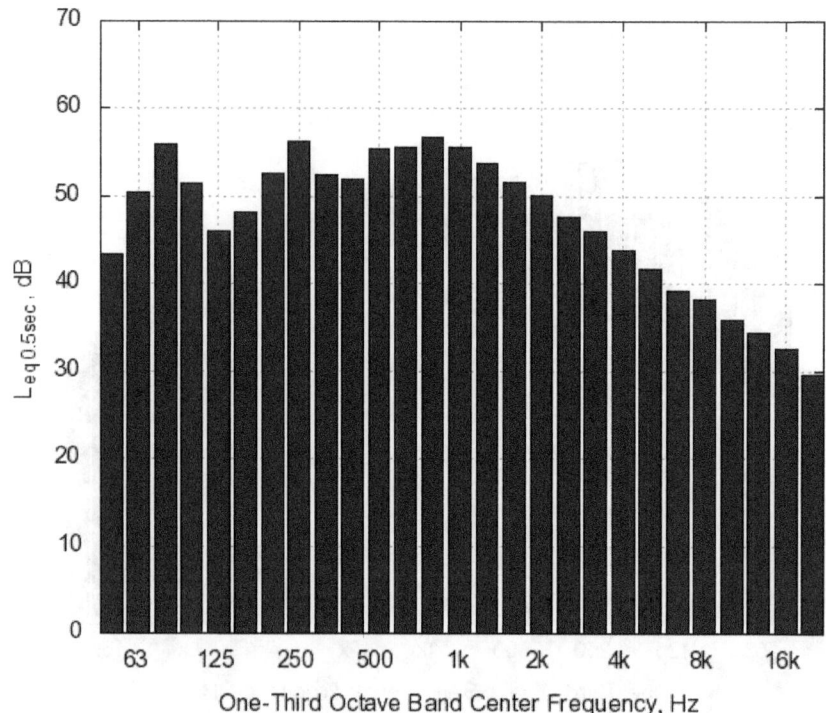

Figure A-67. Matrix One-Third Octave Band Levels for Acceleration Passby at 12 ft Microphone

Appendix A: Acoustic Data for Vehicles

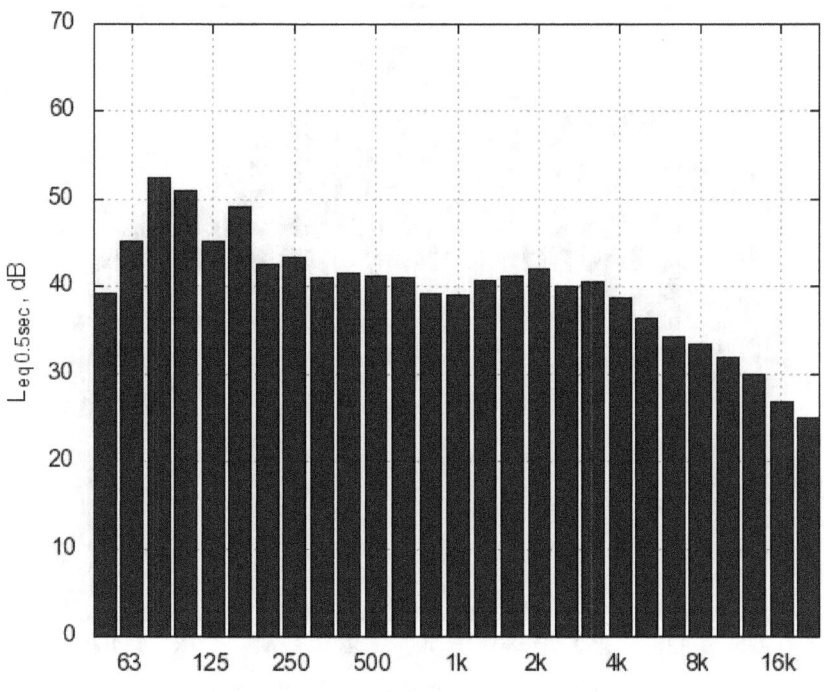

Figure A-68. Matrix One-Third Octave Band Levels for Reverse 5 mph Constant Speed Passby at 12 ft Microphone

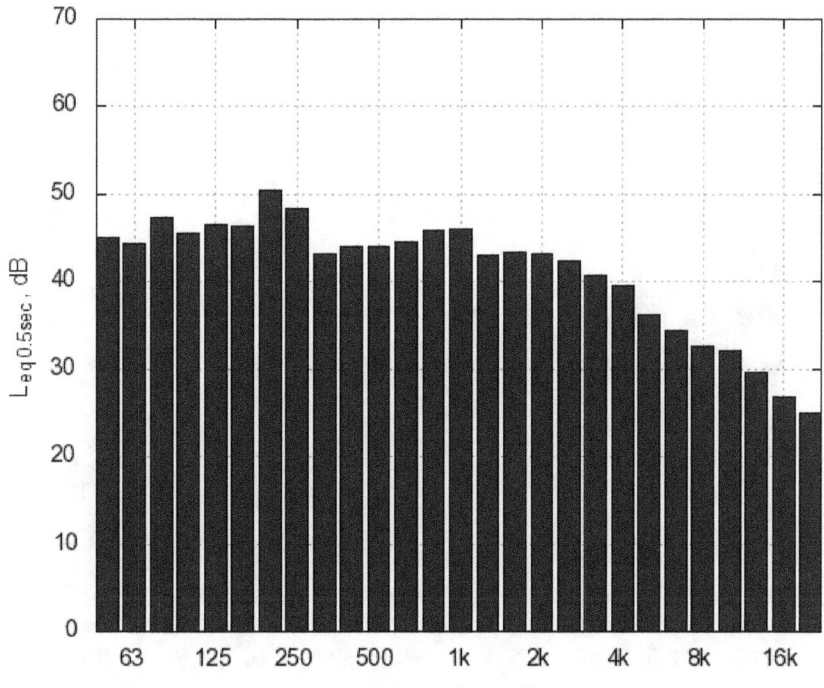

Figure A-69. Matrix One-Third Octave Band Levels for Deceleration Passby at 12 ft Microphone

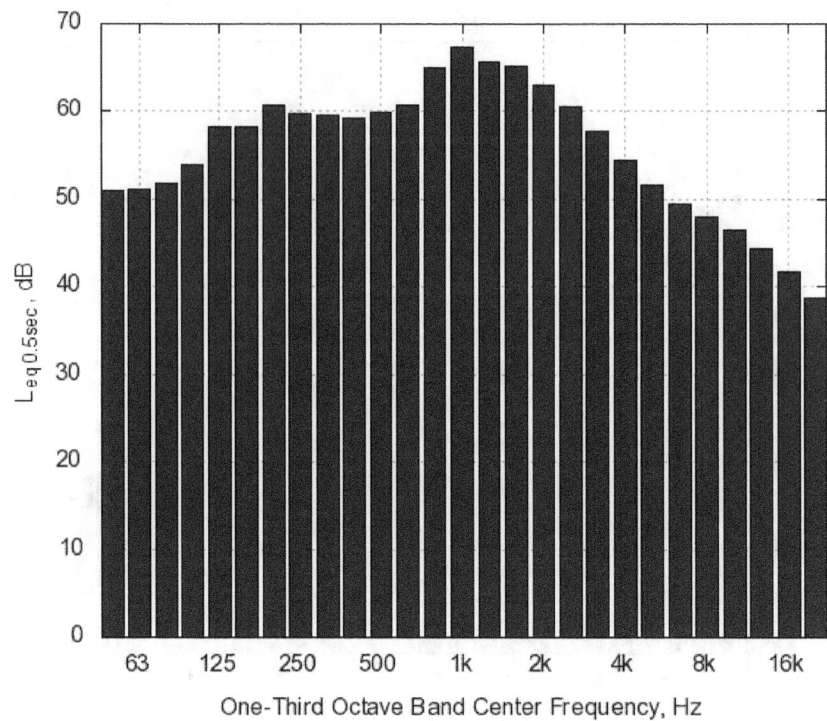

Figure A-70. Matrix One-Third Octave Band Levels for 40 mph Constant Speed Passby at 12 ft Microphone

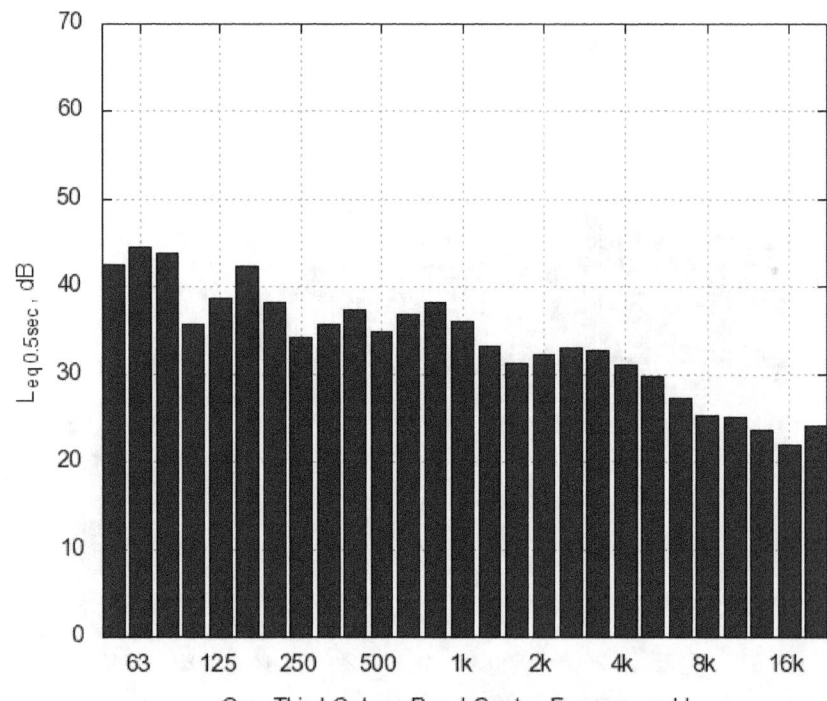

Figure A-71. Honda Civic Hybrid One-Third Octave Band Levels for Idle at 12 ft Microphone

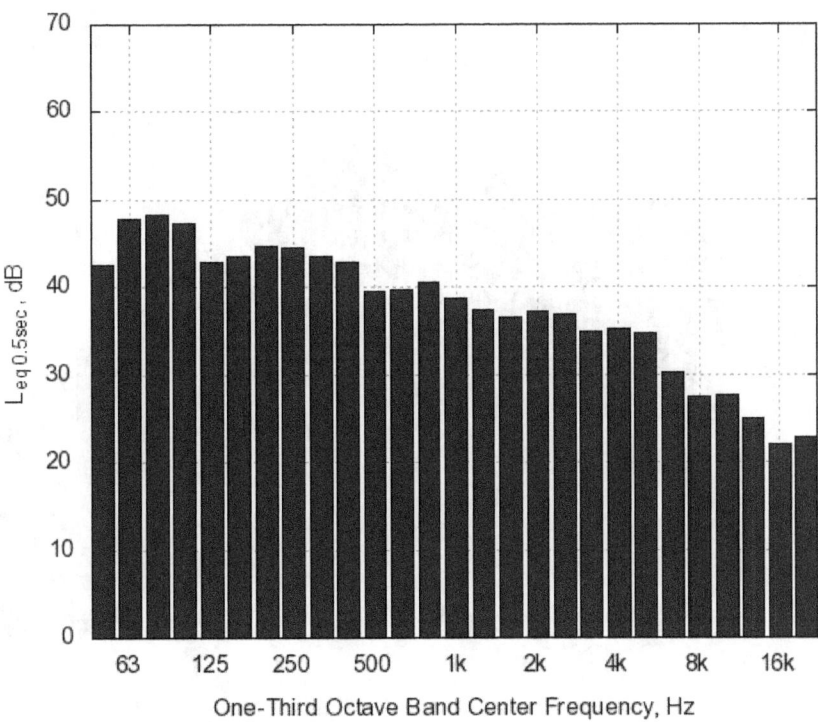

Figure A-72. Honda Civic Hybrid One-Third Octave Band Levels for 6 mph Constant Speed Passby at 12 ft Microphone

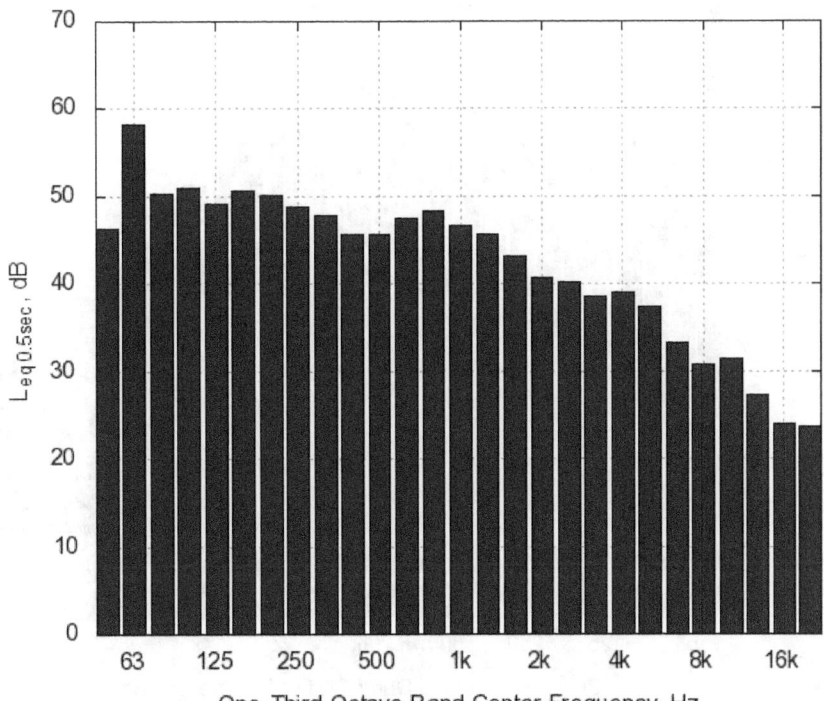

Figure A-73. Honda Civic Hybrid One-Third Octave Band Levels for 10 mph Constant Speed Passby at 12 ft Microphone

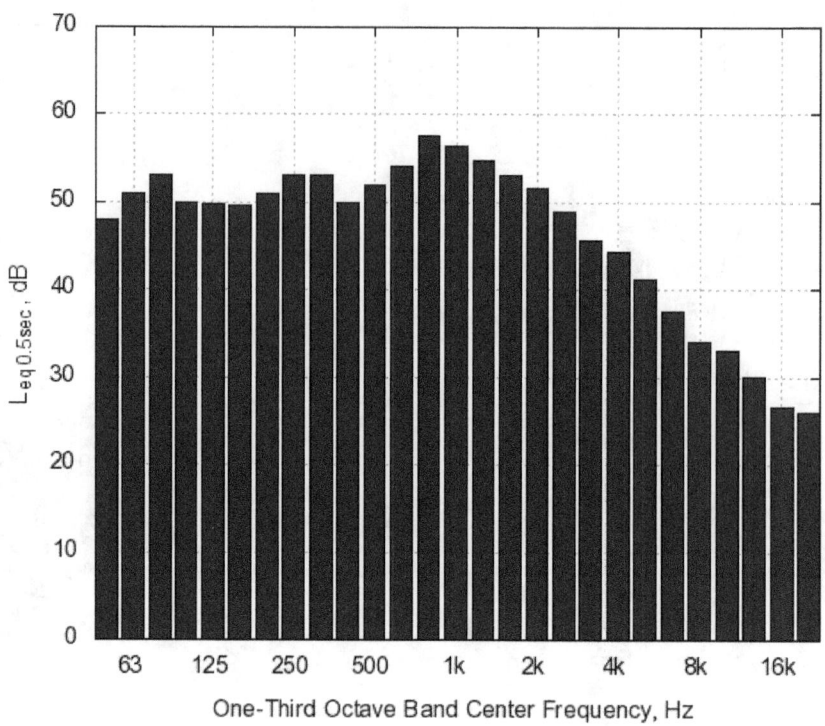

Figure A-74. Honda Civic Hybrid One-Third Octave Band Levels for 20 mph Constant Speed Passby at 12 ft Microphone

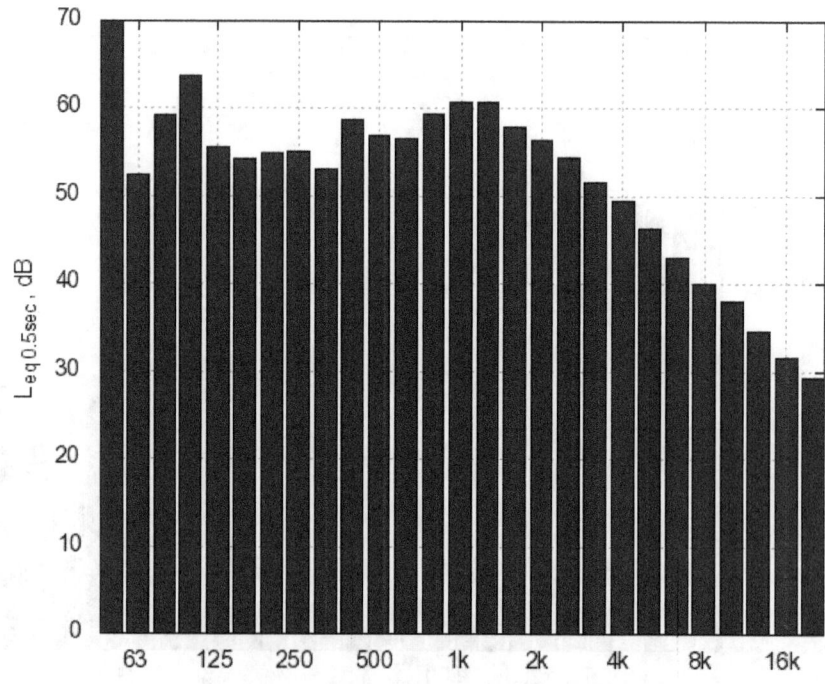

Figure A-75. Honda Civic Hybrid One-Third Octave Band Levels for 30 mph Constant Speed Passby at 12 ft Microphone

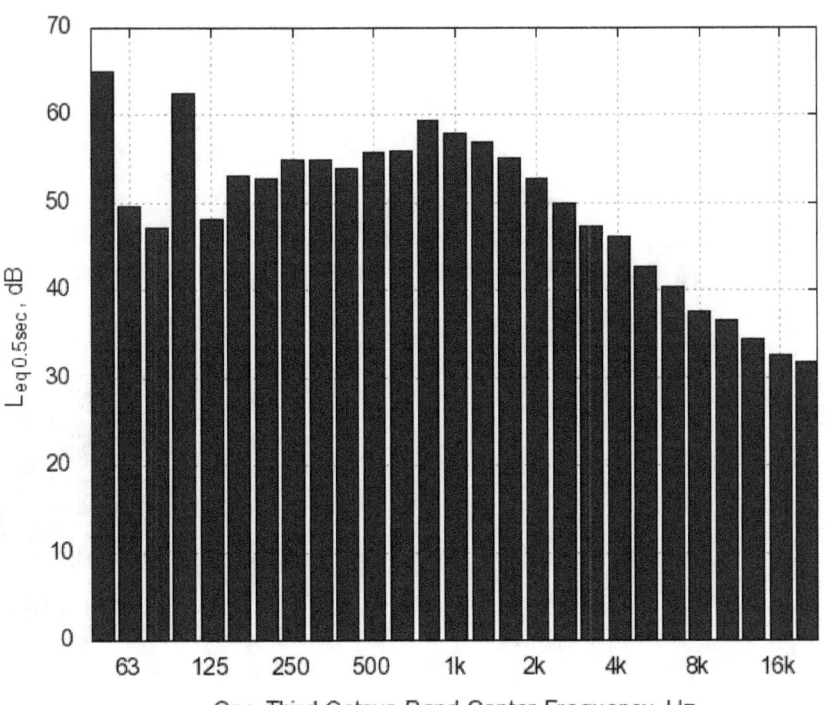

-Figure A-76. Honda Civic Hybrid One-Third Octave Band Levels for Acceleration Passby at 12 ft Microphone

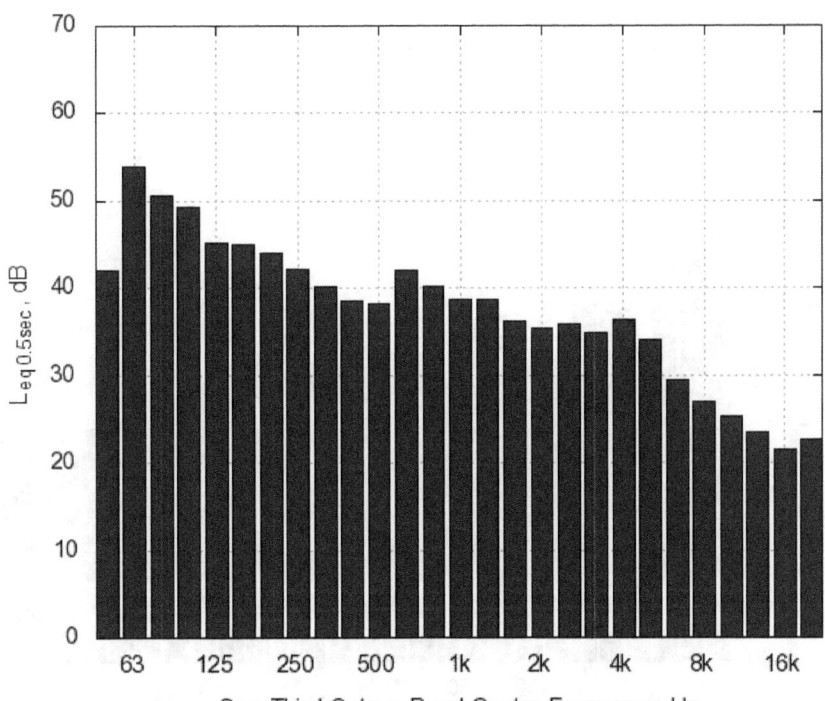

Figure A-77. Honda Civic Hybrid One-Third Octave Band Levels for Reverse 5 mph Constant Speed Passby at 12 ft Microphone

Appendix A: Acoustic Data for Vehicles

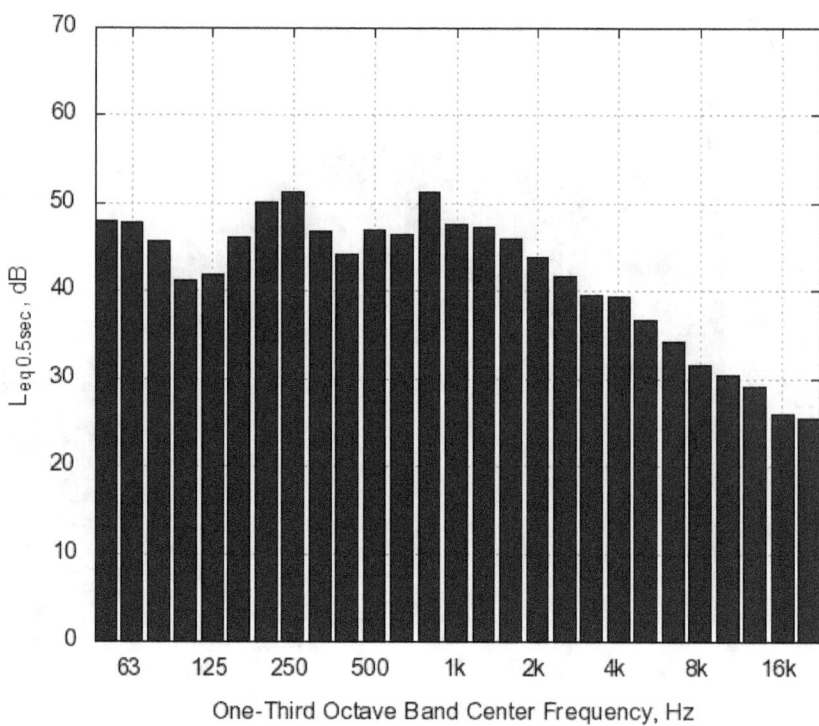

Figure A-78. Honda Civic Hybrid One-Third Octave Band Levels for Deceleration Passby at 12 ft Microphone

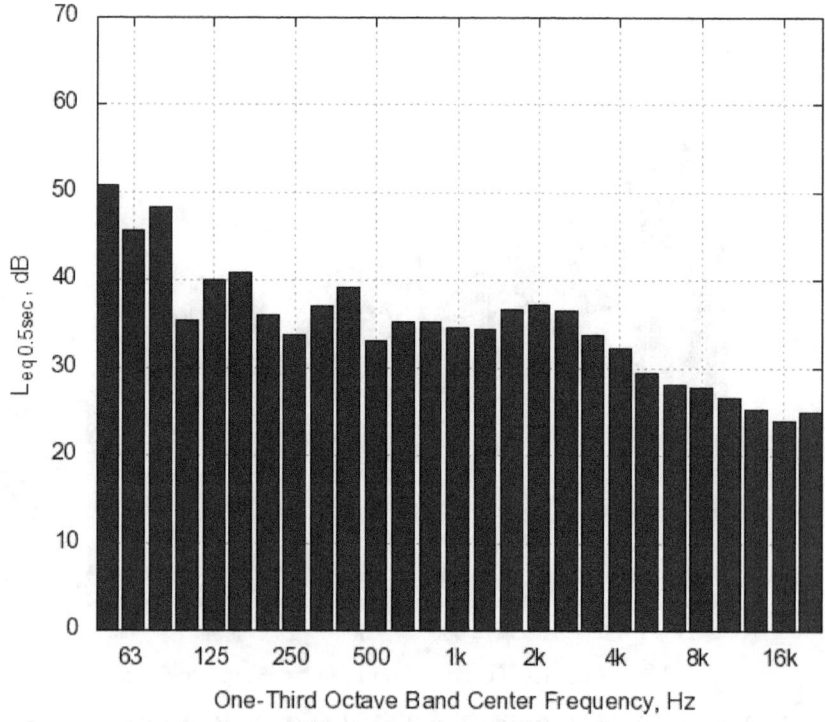

Figure A-79. Honda Civic ICE One-Third Octave Band Levels for Idle at 12 ft Microphone

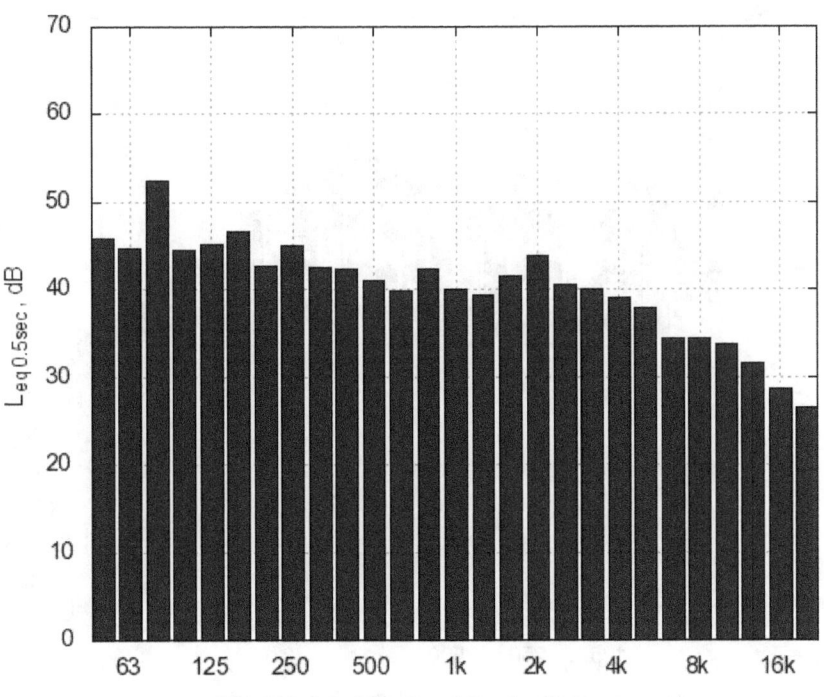

Figure A-80. Honda Civic ICE One-Third Octave Band Levels for 6 mph Constant Speed Passby at 12 ft Microphone

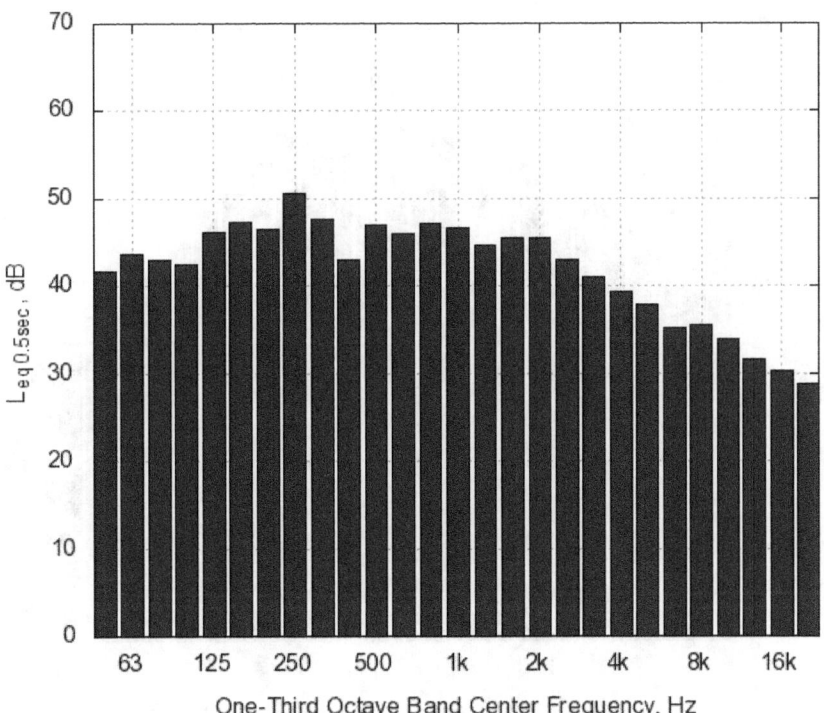

Figure A-81. Honda Civic ICE One-Third Octave Band Levels for 10 mph Constant Speed Passby at 12 ft Microphone

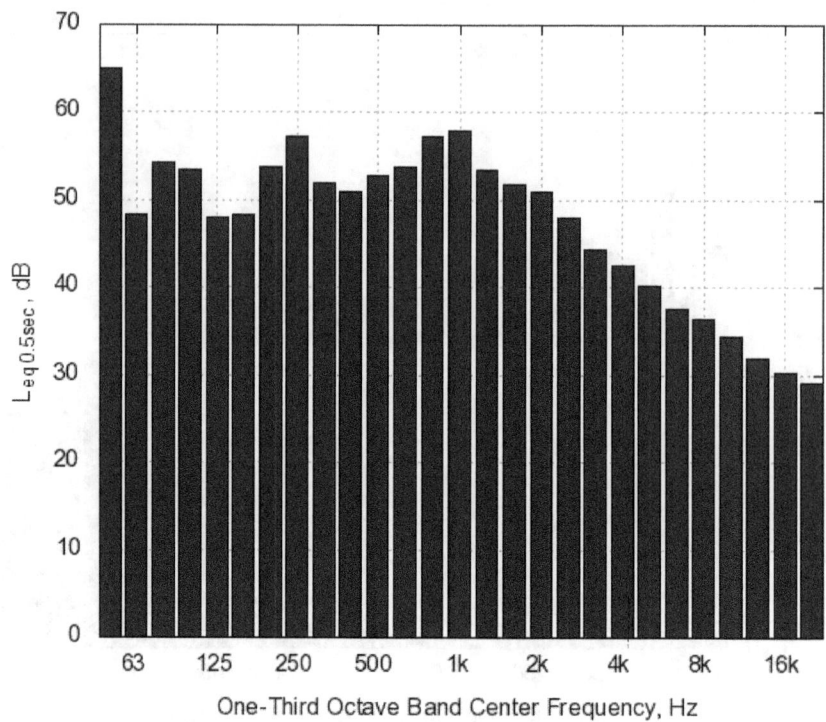

Figure A-82. Honda Civic ICE One-Third Octave Band Levels for 20 mph Constant Speed Passby at 12 ft Microphone

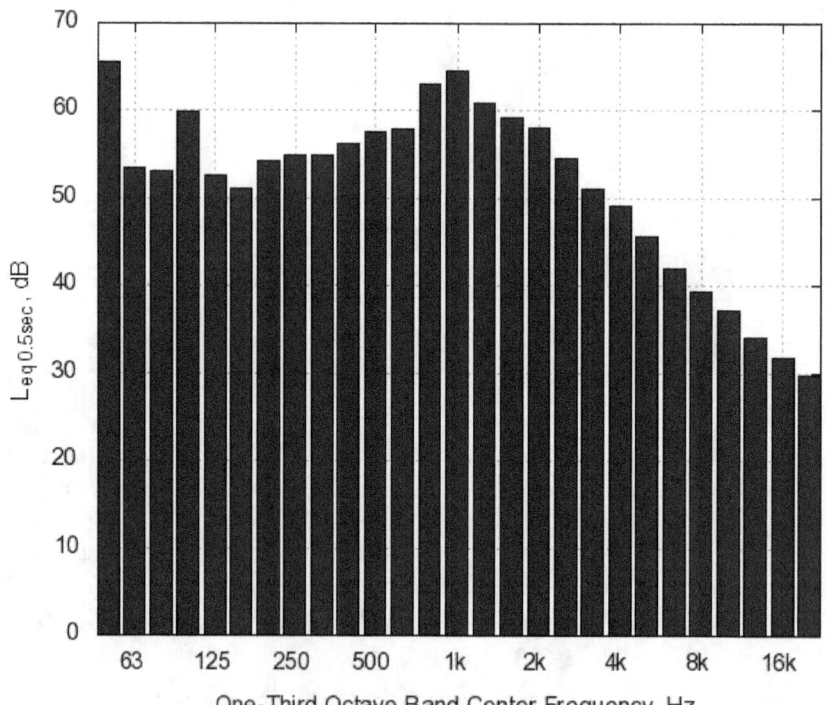

Figure A-83. Honda Civic ICE One-Third Octave Band Levels for 30 mph Constant Speed Passby at 12 ft Microphone

Appendix A: Acoustic Data for Vehicles

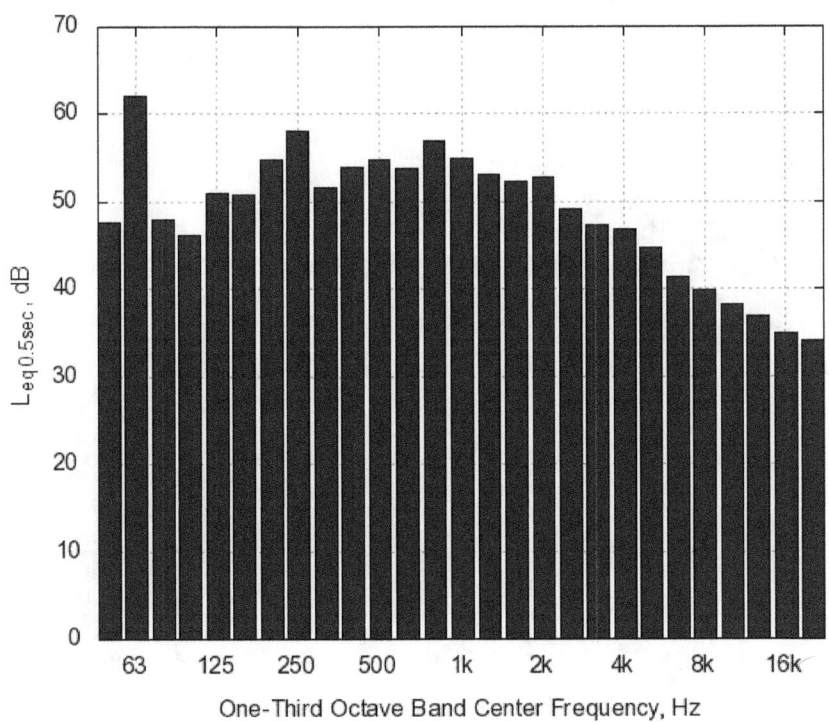

Figure A-84. Honda Civic ICE One-Third Octave Band Levels for Acceleration Passby at 12 ft Microphone

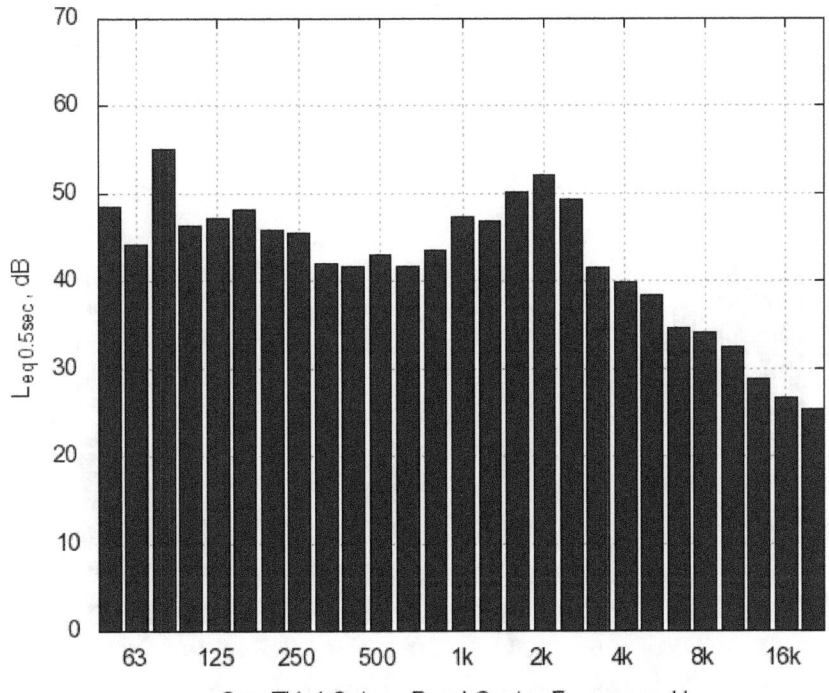

Figure A-85. Honda Civic ICE One-Third Octave Band Levels for Reverse 5 mph Constant Speed Passby at 12 ft Microphone

Appendix A: Acoustic Data for Vehicles

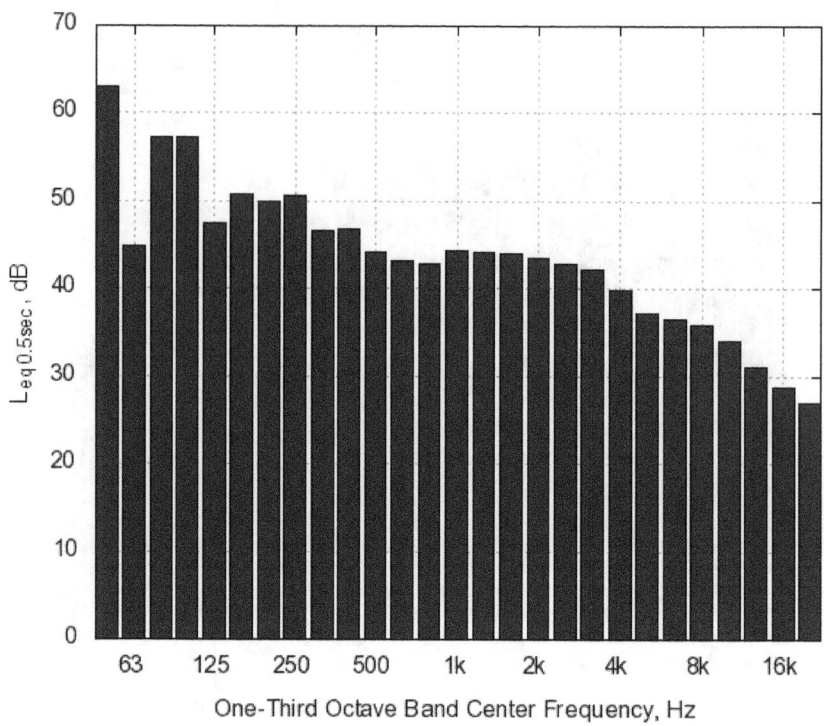

Figure A-86. Honda Civic ICE One-Third Octave Band Levels for Deceleration Passby at 12 ft Microphone

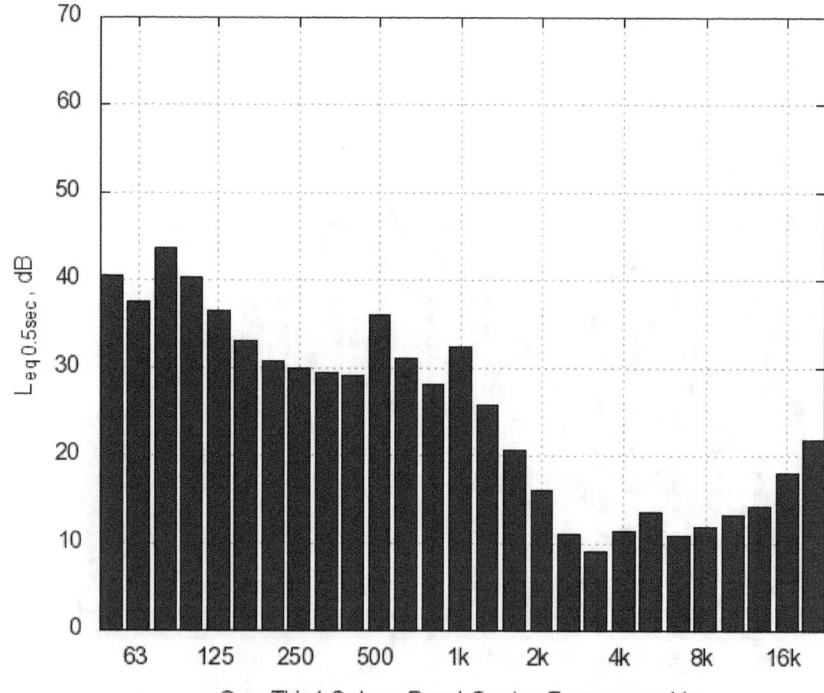

Figure A-87. Toyota Highlander Hybrid One-Third Octave Band Levels for Idle at 12 ft Microphone

Appendix A: Acoustic Data for Vehicles

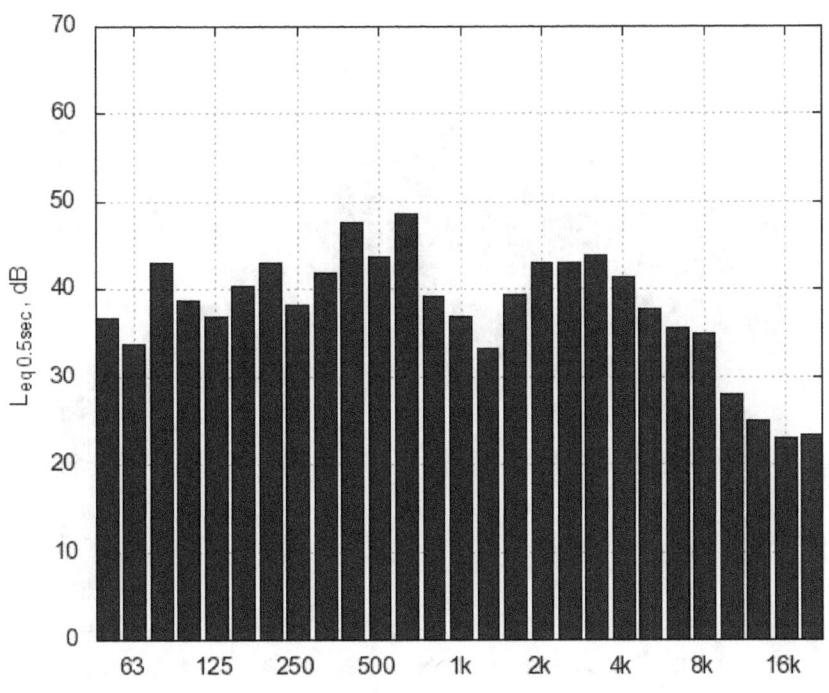

Figure A-88. Toyota Highlander Hybrid One-Third Octave Band Levels for 6 mph Constant Speed Passby at 12 ft Microphone

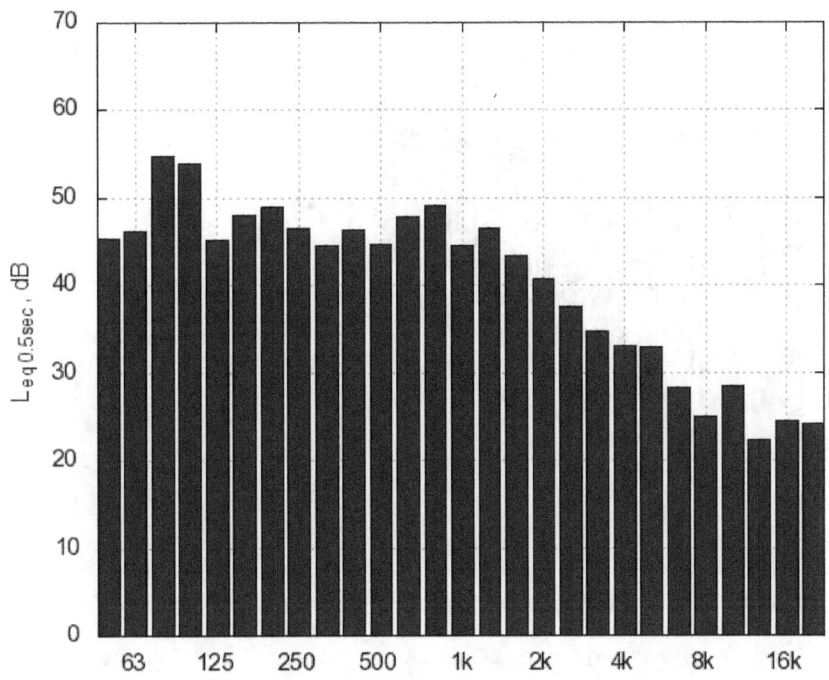

Figure A-89. Toyota Highlander Hybrid One-Third Octave Band Levels for 10 mph Constant Speed Passby at 12 ft Microphone

Appendix A: Acoustic Data for Vehicles

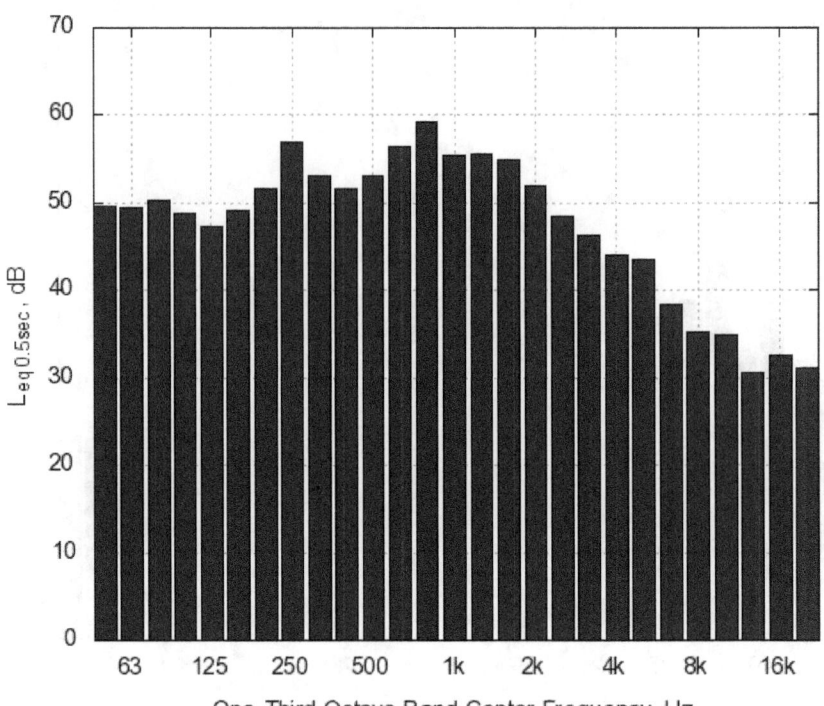

Figure A-90. Toyota Highlander Hybrid One-Third Octave Band Levels for 20 mph Constant Speed Passby at 12 ft Microphone

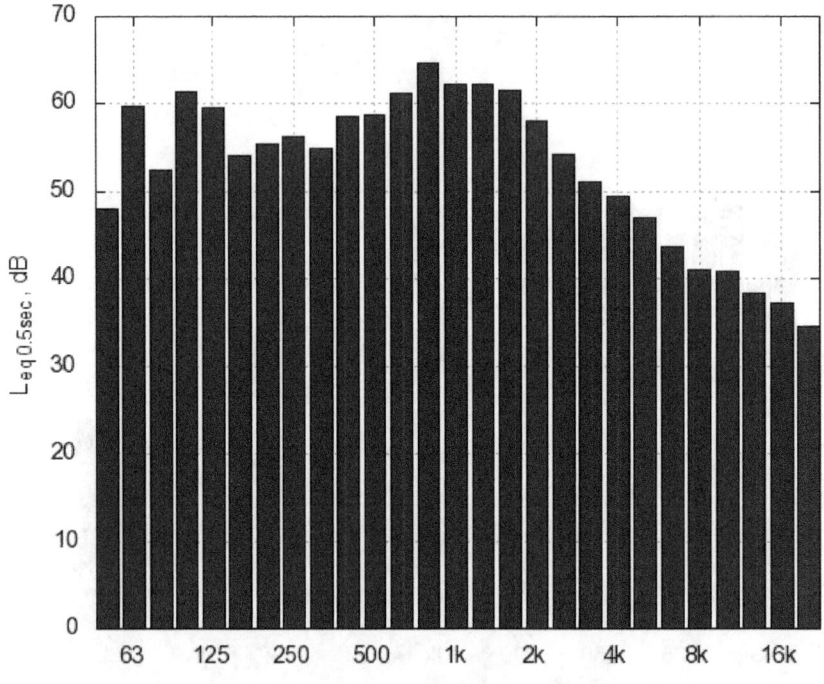

Figure A-91. Toyota Highlander Hybrid One-Third Octave Band Levels for 30 mph Constant Speed Passby at 12 ft Microphone

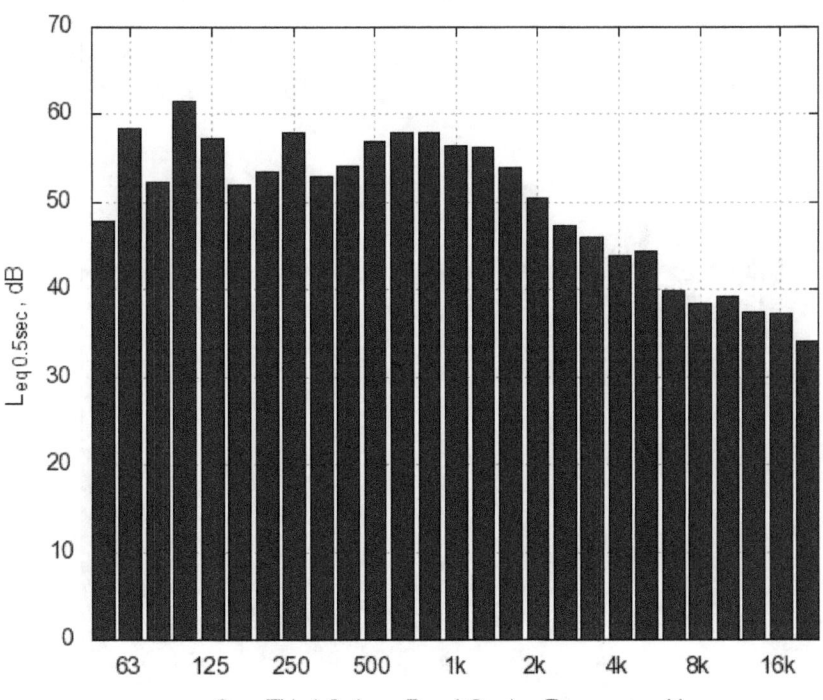

Figure A-92. Toyota Highlander Hybrid One-Third Octave Band Levels for Acceleration Passby at 12 ft Microphone

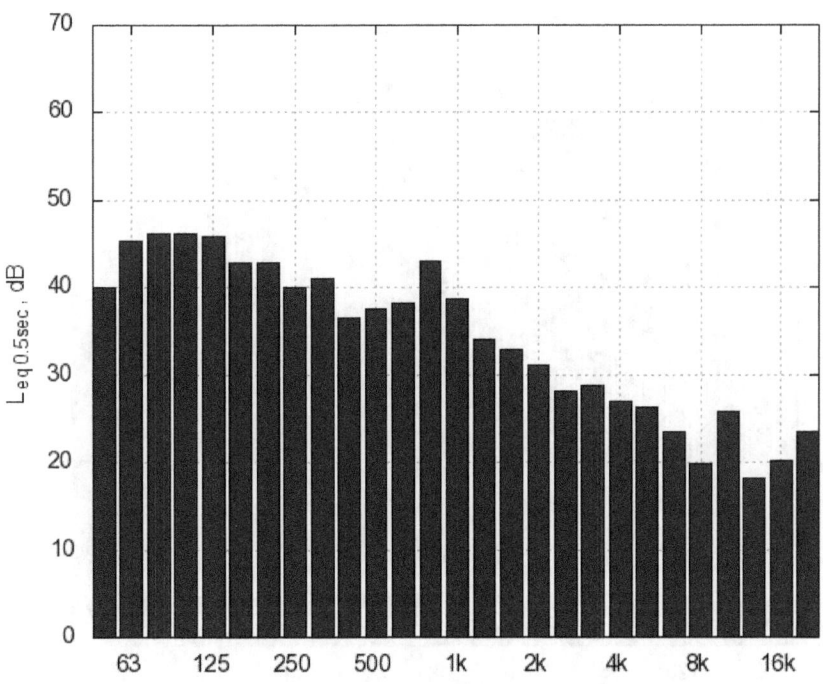

Figure A-93. Toyota Highlander Hybrid One-Third Octave Band Levels for Reverse 5 mph Constant Speed Passby at 12 ft Microphone

Appendix A: Acoustic Data for Vehicles

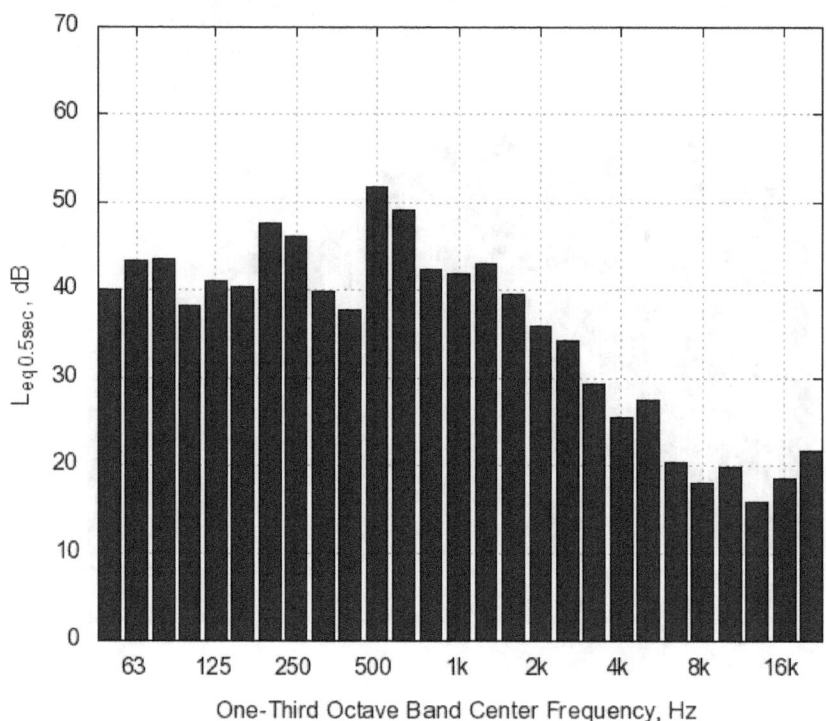

Figure A-94. Toyota Highlander Hybrid One-Third Octave Band Levels for Deceleration Passby at 12 ft Microphone

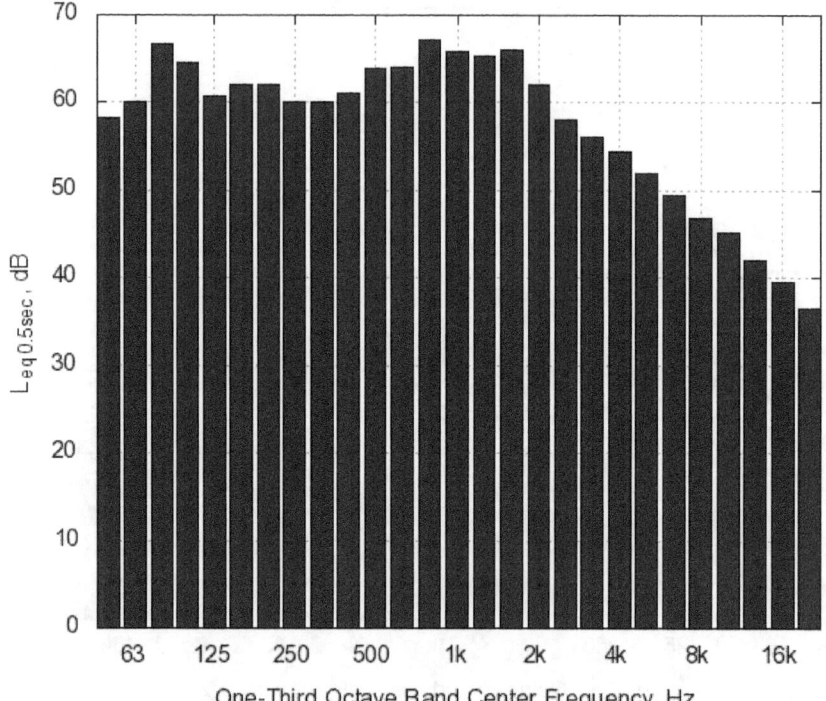

Figure A-95. Toyota Highlander Hybrid One-Third Octave Band Levels for 40 mph Constant Speed Passby at 12 ft Microphone

Appendix A: Acoustic Data for Vehicles

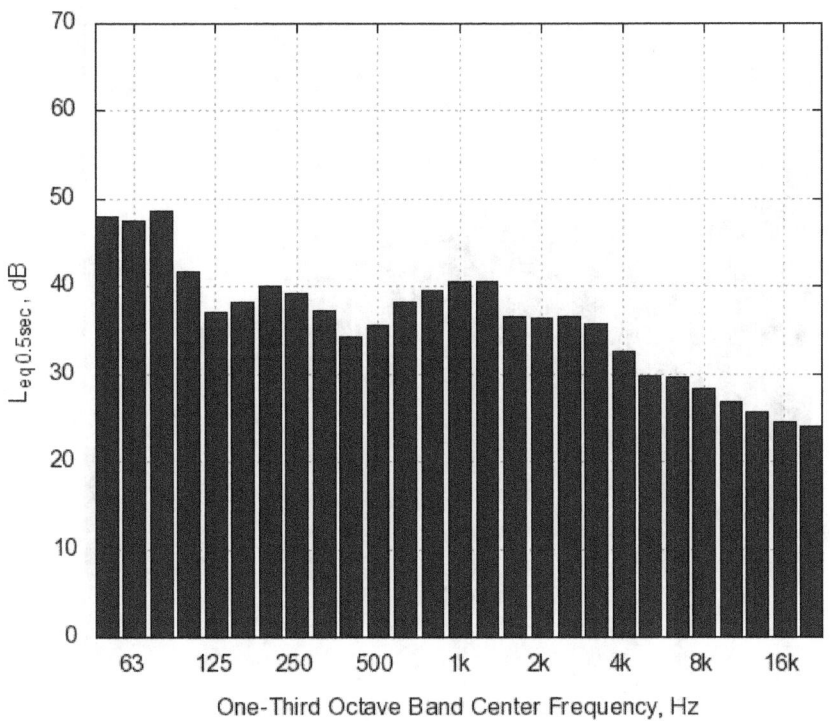

Figure A-96. Toyota Highlander ICE One-Third Octave Band Levels for Idle at 12 ft Microphone

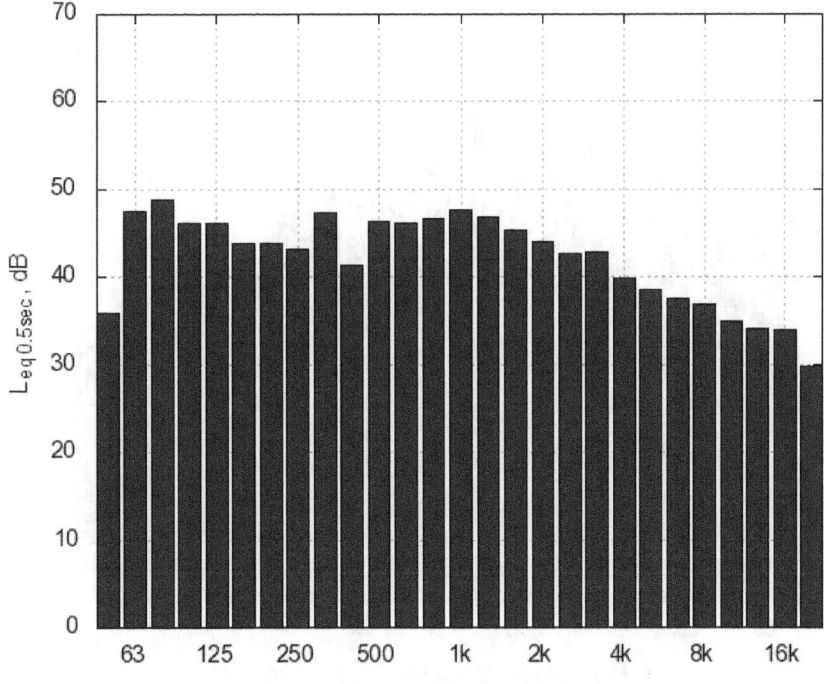

Figure A-97. Toyota Highlander ICE One-Third Octave Band Levels for 6 mph Constant Speed Passby at 12 ft Microphone

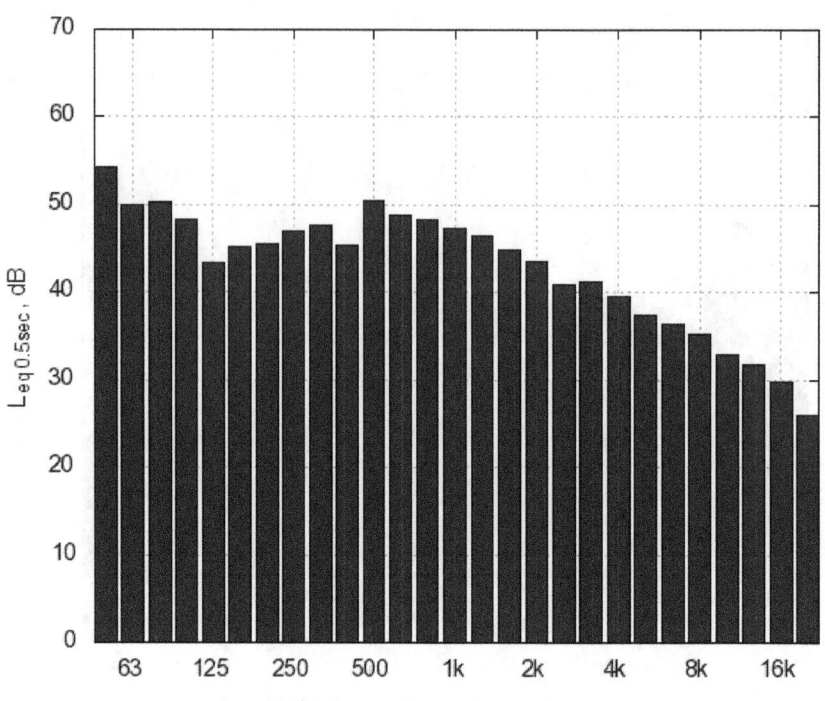

Figure A-98. Toyota Highlander ICE One-Third Octave Band Levels for 10 mph Constant Speed Passby at 12 ft Microphone

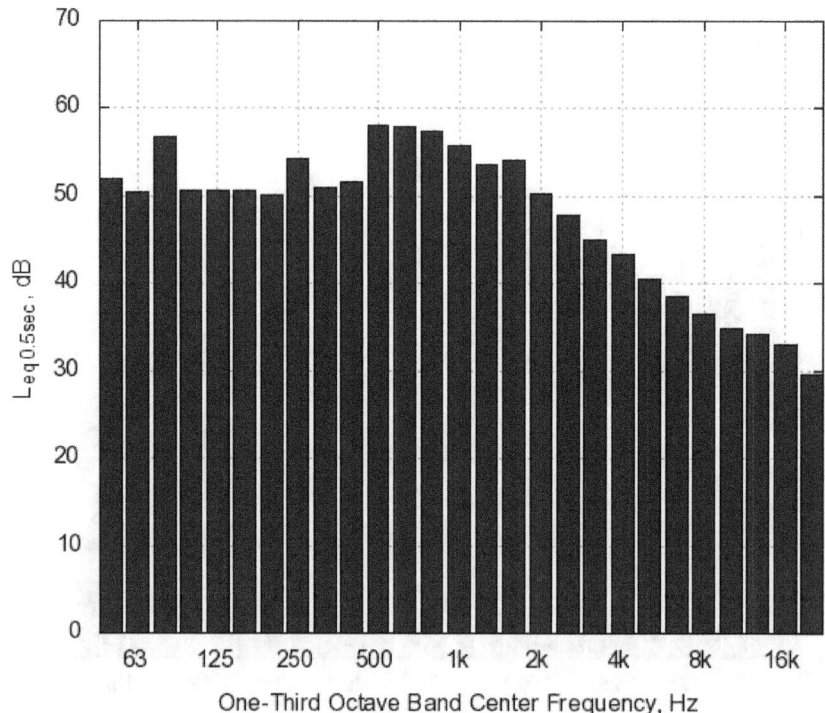

Figure A-99. Toyota Highlander ICE One-Third Octave Band Levels for 20 mph Constant Speed Passby at 12 ft Microphone

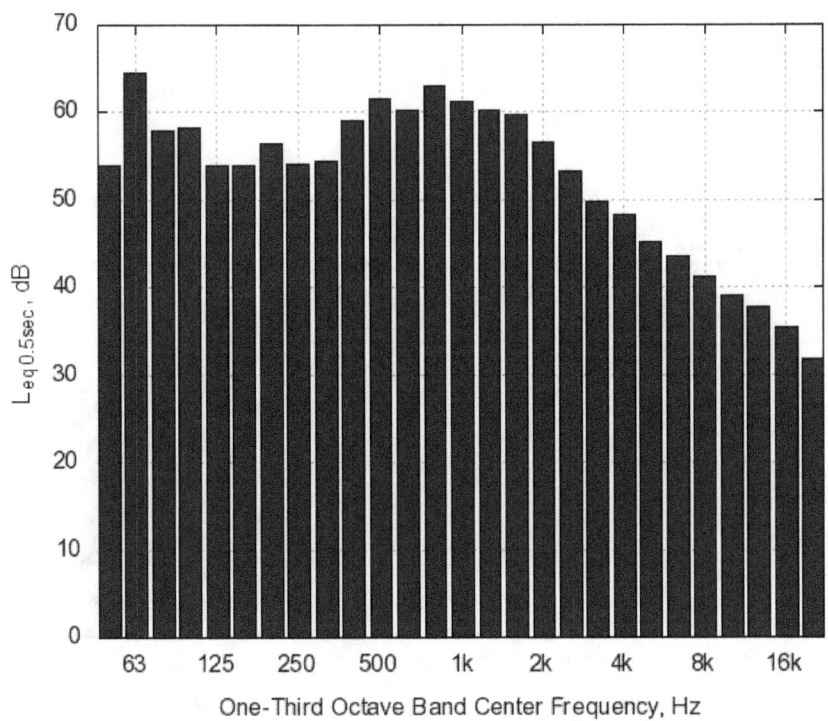

Figure A-100. Toyota Highlander ICE One-Third Octave Band Levels for 30 mph Constant Speed Passby at 12 ft Microphone

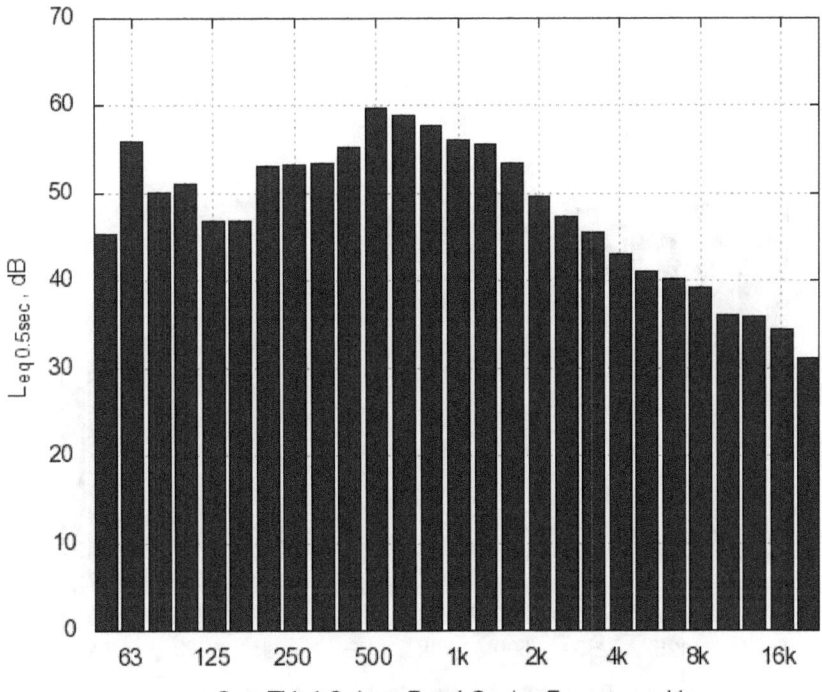

Figure A-101. Toyota Highlander ICE One-Third Octave Band Levels for Acceleration Passby at 12 ft Microphone

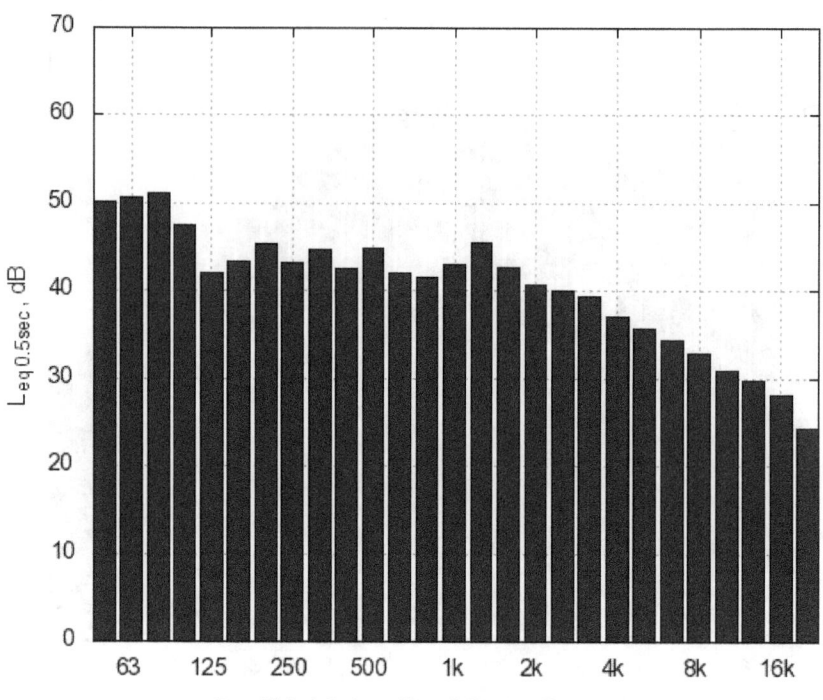

Figure A-102. Toyota Highlander ICE One-Third Octave Band Levels for Reverse 5 mph Constant Speed Passby at 12 ft Microphone

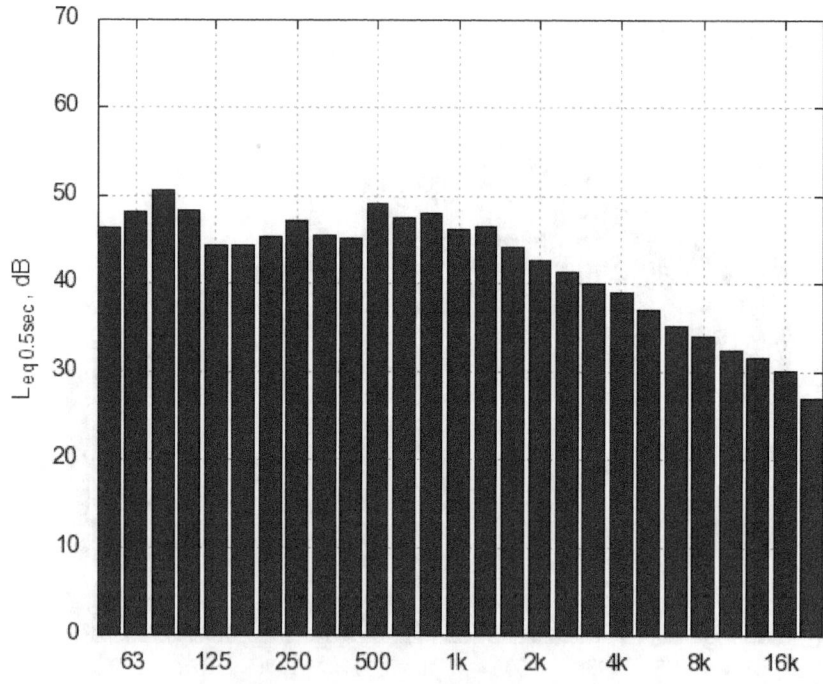

Figure A-103. Toyota Highlander ICE One-Third Octave Band Levels for Deceleration Passby at 12 ft Microphone

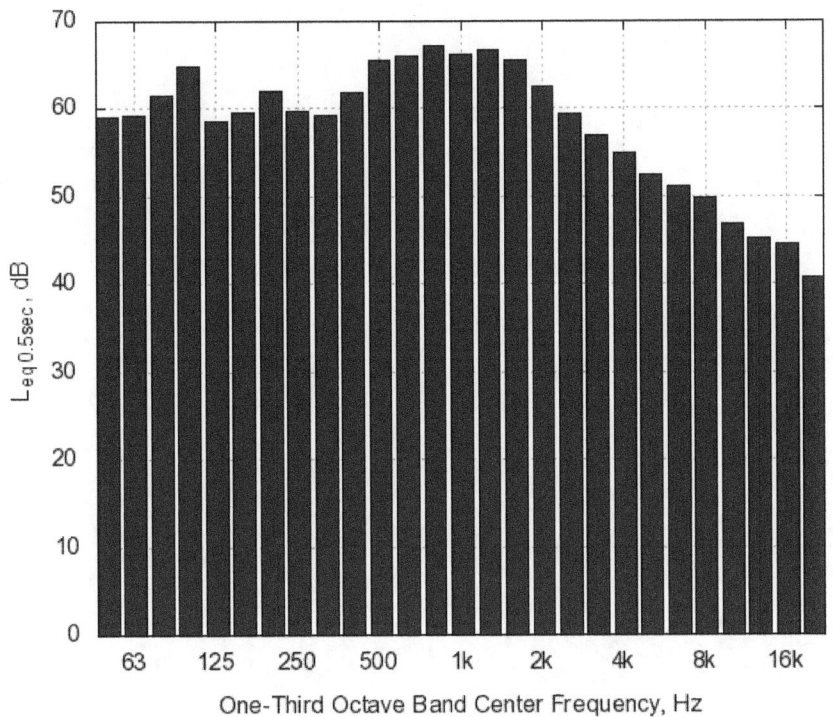

Figure A-104. Toyota Highlander ICE One-Third Octave Band Levels for 40 mph Constant Speed Passby at 12 ft Microphone

DOT HS 811 304
April 2010

U.S. Department
of Transportation
**National Highway
Traffic Safety
Administration**

www.ingramcontent.com/pod-product-compliance
Lightning Source LLC
Chambersburg PA
CBHW080252180526
45167CB00006B/2504